ABOUT BOSTON

Sight, Sound, Flavor & Inflection

Some Other Books by David McCord

VERSE: Floodgate · The Crows · Bay Window Ballads · Twelve Verses from XII Night · And What's More · On Occasion · Far and Few · A Star by Day · Odds without Ends · The Old Bateau and Other Poems · Take Sky

ESSAYS: Oddly Enough · Stirabout · H.T.P.—Portrait of a Critic · Notes on the Harvard Tercentenary · The Camp at Lockjaw · In Sight of Sever · Homage to New England · The Fabrick of Man

ANTHOLOGIES: Once and for All · What Cheer · The Pocket Book of Humorous Verse

About Boston

SIGHT, SOUND, FLAVOR & INFLECTION

by David McCord

WITH DRAWINGS BY THE AUTHOR

"Yes'm, old friends is always best,
'less you can catch a new one that's
fit to make an old one out of."

SARAH ORNE JEWETT

LITTLE, BROWN AND COMPANY, BOSTON, TORONTO

The extract from *The Crows,* by David McCord, copyright, 1934, by Charles Scribner's Sons.

The extract from *The Kid,* by Conrad Aiken, copyright, 1947, by Conrad Aiken; reprinted by permission of Duell, Sloan & Pearce, Inc.

The lines from "Gulls on the Ice," by David McCord, are reprinted by permission from *The Saturday Review of Literature,* April 9, 1938.

The lines from "Early for Once," by David McCord, originally appeared in *The Yale Review,* copyright, 1940, by Yale University Press and reprinted with their permission.

The extract from *The Maritime History of Massachusetts, 1783–1860,* copyright, 1921, by S. E. Morison, reprinted by permission of Houghton Mifflin Company.

The lines from "Immortal Autumn" are from *Poems, 1924–1933,* copyright, 1933, by Archibald MacLeish, reprinted by permission of Houghton Mifflin Company.

The extract from *Travels in Philadelphia,* by Christopher Morley, copyright 1920, 1948, by Christopher Morley, published by J. B. Lippincott Company.

The lines from *And What's More,* by David McCord, copyright, 1941, by David McCord, reprinted by permission of Coward-McCann, Inc.

The extracts from *H. T. P.—Portrait of a Critic,* by David McCord, copyright, 1935, by David McCord, reprinted by permission of Coward-McCann, Inc.

The extracts from *Boston after Bulfinch,* by Walter H. Kilham, copyright, 1946, by The President and Fellows of Harvard College.

The extract from *Collected Poems,* by Conrad Aiken, copyright, 1953, by Conrad Aiken; reprinted by permission of Oxford University Press, Inc.

D

Published simultaneously in Canada
by Little, Brown & Company (Canada) Limited

PRINTED IN THE UNITED STATES OF AMERICA

ACKNOWLEDGMENTS Small as it is, this book has many debts. I am grateful first of all to Robert W. Maynard, Carl N. Schmalz, James C. Davis of R. H. Stearns & Company, and Walton Butterfield and Mrs. Sally Larkin for early interest and critical help in preparing for broadcasting the original manuscript which forms the basis of this book; to Miss Marion L. Anderson, Miss Mary F. MacDougall, and Miss Irene M. Carey for secretarial help and suggestions; to Stephen Burke, Richard Bauer, Wendell Davis, Avner Rakov, and Robert Rissling of Station WBZ, Boston; to Mark A. DeWolfe Howe for many kindnesses and friendly advice; to John H. Bradley, Jr., for geological and meteorological assistance; to George C. Greener, Director of the North Bennet Street Industrial School; to Robert H. Haynes, Assistant Librarian of the College Library, Harvard University, and his Research Assistant, Miss Eleanor Hall; to Walter Muir Whitehill, Director and Librarian of the Boston Athenaeum; to John J. Rowlands of the Massachusetts Institute of Technology; to George E. Judd of Symphony Hall, Boston; to officials of the Boston Public Library, Boston Chamber of Commerce, Boston Museum of Fine Arts; and to my publishers, Ralph A. Beebe and LeBaron R. Barker of Doubleday & Company.

I have also to thank the following authors and publishers for permission to use brief quotations as follows: The editors of the *Harvard Alumni Bulletin* for a verse of my own and for brief quotations from an article by the late Robert DeCourcy Ward which appeared originally in their pages; Conrad Aiken and Duell, Sloan & Pearce, Inc., for four stanzas from *The Kid;* Charles Scribner's Sons for eight lines by the author from "The Crows" and "8:15" published in *The Crows; The Saturday Review of Literature* for four stanzas from "Gulls on the Ice" by the author; *The Yale Review* for "Early for Once" by the author; Walter H. Kilham and the Harvard University Press for brief quotations from and general pillaging of *Boston after Bulfinch;* Coward-McCann, Inc., for quotations and material of my own from *H. T. P.—Portrait of a Critic* and for six lines of my own from *And What's More;* Samuel E. Morison and Houghton Mifflin Company for two brief passages from *The Maritime History of Massachusetts;* Archibald MacLeish and Houghton Mifflin Company for two stanzas from "Immortal Autumn," *Poems, 1924–1933;* Christopher Morley and J. B. Lippincott Company for a brief quotation from *Travels in Philadelphia* by Christopher Morley.

D. T. W. McC.

PREFACE

Somehow the thought of another book about Boston is like the thought of another book about Shakespeare. What more is there to say? Nothing and everything. Let me tell you what Francis Parkman set down in his recently published *Journals:* "I like the Scotch–I like the country and everything in it." A big statement issuing from a short visit. On his arrival at Edinburgh by night he added: "The view from Calton Hill is, to my way of thinking, the only city view I ever saw that deserves to be called sublime." A big word in the wake of a single glance. And yet it is the right word, as all of us who have seen it after him will agree.

I have tried to look at Boston much as Parkman, a Bostonian, looked at Edinburgh. I am not a Bostonian by birth, and my glance must necessarily be a sweeping one. As a native of New York, I am looking at a smaller town; as an Oregonian raised and schooled, I am looking at a large and ancient metropolis. I have found the dual vantage point to be helpful, and I trust I have not failed to see the city for the trees.

This book grew out of a series of brief broadcasts which I was giving not so many months ago. I have somewhat revised, expanded, and rearranged them for a possible audience of readers. Unlike Max Beerbohm, I believe that all writing should be for the ear as well as for the eye. He was at some pains to explain that *Mainly on the Air* (1946) was in part "composed for the ears of listeners," but it will take a sensitive reader indeed to find the prose of the incomparable Sir Max of today differing from or inferior to the prose of the young writer just down from Oxford in the middle nineties. So I cannot charge the faults of these sketches to the original medium for which they were written. Perhaps it is too much to call them sketches: talks is a better name for them. But one thing should be made clear. I am dealing throughout with a series of views, and the marginal notes on these views are chiefly concerned with sight, sound, color, flavor, inflection, and atmospherics. I am not here concerned with city government or dirty streets or book-banning or the pips of cheapness that any radar can detect within the city limits. You would not stand in front of a fine old painting of character and say that it might be good if it had a new frame and a cleaning and a coat of varnish. I have tried, merely, to see what Time as an artist has given us.

Even so, I am handicapped by many things. I hear the voices of competent people who have found Boston to be everything from the sublime to the ridiculous, from a Mecca to a museum, from a living force to a city of the dead. History and literature can offer evidence in support of every attitude. There is, for instance, Emersonian veneration. In

Dr. Holmes's day, as Van Wyck Brooks has pointed out, "Boston had earned its good name; and, as for the unkind things that were said about it, one usually found some Boston man who had said them first and better." At the close of *New England: Indian Summer* the reader should not be surprised to find many Boston men saying them first, better—and bitter. That Boston in Dr. Holmes's day "had no second as a seat of culture" does not mean that the place could not be anathema too. "I hate Boston," said Lincoln Steffens. "I don't know why. . . . The general spirit is so far, far, far back that it gets on my nerves." Charles Francis Adams felt that "there is no current of fresh outside life everlastingly flowing in and passing out." I remember a letter that Elmer Davis quoted in an article on Boston in *Harper's* back in January, 1928: "Boston is going to hang on to the ideals that built it, just as long as they can be hung on to. It may be a futile struggle but it is gallant and pathetic." All this—and the voices are endless—is directed, it seems to me, at what Howells in 1884 described in *The Rise of Silas Lapham* as "a society where Middlesexes have married Essexes and produced Suffolks for two hundred and fifty years." The topside one-sidedness of Boston is not what interests me. I prefer a phrase of Henry Adams, though not exactly in the Boston sense that Adams used it: "a peculiar and perplexing amalgam." I prefer to do what Mr. Nathaniel Alden in *The Last Puritan* did not do. "Never, except to funerals, did Mr. Nathaniel Alden walk *down* Beacon Hill." But that, I trust, is not my errand.

One thing more. Dos Passos speaks somewhere with his telescoped words about a *steelengraving.*

In my chapters on architecture, the islands, and geology I have leaned so heavily on my acknowledged sources that it seems to me I have effected in some respects a stolengraving. For this an unqualified layman asks forgiveness. The picture, like the moon, had to be gibbous if not round. Last of all, should this appear in the end to be a study in quiescence, be of good cheer none the less. Out of Greater Boston in a time of war came useful things like the fractionation of blood, oxygen masks for high-altitude flying, counter radar, synthetic quinine, underwater sound equipment, G.I. manuals on edible and poisonous plants in the Pacific, the Navy's huge mechanical computator, and some rather dangerous and destructive weapons.

I shall now put on my rose-colored Polaroid glasses—a Cambridge product.

PREFACE 1973 These few words to usher in the tenth printing of *About Boston* will make little sense unless you read the original preface first. Having just reread it myself, I am inclined to extend it by a couple of paragraphs before I come to the crux of this new one.

To begin with, I did not say in 1948 that the book manuscript had been turned down by Houghton Mifflin Company and by Little, Brown and Company, both of Boston, before it was accepted by Doubleday in New York; accepted, I should add, largely because some youngish female reader had said most kindly: "This sounds like Emerson advertising Carter's Little Liver Pills, and I think we had better take it." Doubleday reprinted *About Boston* four times; and then, like a fresh-run salmon swimming up her unpolluted river to the spawning grounds, it rested for a while in a quiet pool of possible oblivion, until my present publisher was persuaded (not by me) to purchase the plates from Doubleday for an undisclosed sum—three small figures, I should guess—and issue an abridged edition of eleven chapters, with my Beacon Hill watercolor on the jacket, for the American Bar Association assembling in the Back Bay. At the same time, the unabridged book was also reissued; but when, in a few years, this printing likewise became exhausted, my anadromous salmon began to look like a played-out fish.

But, no! One dazzling morning I received word by letter (from a dozen blocks away) that *A.B.* really stood for "at bat"; and so, in 1964, the book appeared afresh in hardcover with a new jacket, *and*—first the first time—in paperback. Eight years passed. Then along in December, 1972, I went to my publisher's warehouse to purchase what proved to be the last copy of the paperback edition. The computer was questioned and verified the dismal truth. The computer was praised, and I wished the warehouse a Merry algebraic X. As I write these words, however, my resting but now restive salmon is headed upstream to try the new public fish-ladder just ahead. Bon voyage! Her silver scales reflect the silver anniversary of these pages: 1948–1973.

Comes now the ready question: why have I not revised and enlarged the book? Auden continually revises his published poems, as I for one among many poets and critics wish he would not. I look on this small history as a kind of prose poem in praise or explanation of certain aspects of an old and much-loved city which was, and is struggling to remain, a work of art. I am not a politician, social scientist, behavioral scientist, transportation expert, city planner, architect, or historian of the wrecking ball and parking lot. If there are possible commercial reasons why this book should have its face lifted, aesthetic reasons—which my publisher has advanced ahead of me—forbid it.

A quarter century has seen a lot of changes in every sizable American city. Here in our New England midst, like an old asparagus bed retilled and fertilized, the first few sudden stalks have multiplied both uptown (west) and downtown (east), so that the older, historic Boston, particularly Beacon Hill and the Back Bay, are now but downstage characters against a very modern sky-scraping backdrop; and I shall be the last to say that the effect in certain light and weather is at all unpleasing. For a while, for nearly twenty years in fact, these sudden, doubtful, epigene eruptions shooting skyward looked unpleasant. Prudential Center was a lonely stalagmite at first. Today it is the center of a nearly completed shadow-casting island, though not by a long shot so logically requited by buildings at a distance as the best of lower Montreal or even the Golden Triangle of Pittsburgh. Down State Street way the tall confusion overpowers. It is only the Custom House Tower, sliding in and out of view, that still does sentry duty. The finest accents in a lofty company are those gold hands on the clock.

Such cataloguing in the realm of urban renewal could go on, of course; but most of it would prove quite uncongenial to what this little book is all about. At a time when even university architecture is reaching for the sky, disaster beckons visibly in "the dreadful sundry of the world"—six words from Wallace Stevens. And, since I am quoting him, it is not—or not as yet—"the amassing harmony" which stops me in my tracks. The current biblical dropouts from the King James versionists may presently resolve the awesome truth: that in my Father's House are many high-rise condominiums, apartments, and motels. The daily heartbeat of most cities coast to coast reflects our heparin-thin blood. We use a shrill vocabulary, some of which was not invented when I wrote about the city of my choice. No: to talk of the new in terms of the old would alter text and tempo of these pages. It would be too much like trying out a souped-up engine under the hood of a surviving Model T.

But as to some obvious blind spots in the original text: I had failed (as the broadcasts ended) to come full circle. I did not reach a self-appointed moment to pay my special tribute to the special wonder of Mrs. Jack Gardner's museum out in the Fenway; enchanting then and now for qualities that endeared me as a youngster to the Luxembourg in Paris. I wrote, alas, too early for a chance to do even potential justice to Bradford Washburn's incredible Museum of Science—one measure of man's handiwork to date in the space lab of his own backyard. I likewise regret my failure to mention Chiang Yee's favorite scholar tree in Dr. Holmes's Garden. And in the chapter titled "Aimless" I would have cited Alfred Kazin's fascinating ambiance in *A Walker in the City*—but it was not published until 1951. Opposing the now existing garage under the Common was wholly stupid. It has done no harm. I should have opposed the building of more cars with which to fill it. What city will ever learn that man can enlarge the arteries of entry but that he *cannot* enlarge or increase the number of veins within?

In holding to the old and merely glancing at the

new, I must not lose this chance to mention Philip Johnson's architectural triumph in the huge simplicity of his addition to the Boston Public Library—McKim's now almost eighty-years-old Renaissance masterpiece in Copley Square: a daring modern concept which provides, with unobtrusive top efficiency, a paradise for reader, scholar, writer, teacher, student, lecturer, listener, borrower, and browser. And by so doing it has in fact happily released the central building—perfection as art but ingeniously imperfect as an intended library—to become the city's treasure house of rare books, prints, collections, manuscripts, letters, archives, graphic arts, and the 10 Downing Street of the Commonwealth of Books.

Now last, as a footnote to the chapter "Out of Print," may I unreservedly recommend an emerging classic in the field of landmarks and contours come and gone in Boston? This is Walter Muir Whitehill's *Boston: A Topographical History,* which *is* up to date as well as richly salted with a splendid choice of illustrations past and present. Another volume I would now add to my shelf of books about Boston has the engaging title of *Hidden Gardens of Beacon Hill,* published recently by the Beacon Hill Garden Club. The visitor may never glimpse more than one or two of these retreats *in situ,* but once having seen the photographs he is not likely to forget their teasing existence.

At this abrasive hour in the history of cities, nations, and populace, some words of William Cowper, perhaps the foremost poet of England from 1785 to the close of the century, seem not only comforting, but (on careful examination) most curiously prophetic. He is speaking, in "The Task," of a metropolis. London, surely? For the moment he is speaking of Boston, and to you and me. His apostrophe:

> Oh thou, resort and mart of all the earth,
> Chequer'd with all complexions of mankind,
> And spotted with all crimes; in whom I see
> Much that I love, and more that I admire.

Some readers may recall the first two lines of the final quatrain in Whittier's "My Triumph":

> And for the things I see
> I trust the things to be.

To which I cheerfully subjoin in advance of the Nation's Bicentennial:

> But from your pad, now, Dad,
> Don't things look awful bad?

ABOUT BOSTON

Sight, Sound, Flavor & Inflection

This is Boreyda! and that square minaret, in the town, is of their great mesjid. I saw, as it were, Jerusalem in the desert! . . . The last upshot sunbeams enlightened the dim clay city in glorious manner, and pierced into that dull pageant of tamarisk trees. I asked my rafîk, 'Where are their palms?' He answered, 'Not in this part, they lie behind yonder great dune towards the Wady er-Rummah.'

ARABIA DESERTA

RELIC AND TYPE A city never seems quite so young as it does in the spring, or quite so old as it does in the winter. At the turning point of the seasons Boston is at once both old and new; and perhaps she won't mind if we stop for a moment to look at her. Why should she mind? The country, and indeed the world, has looked at her with mingled admiration, respect, affection, and alarm for something more than three centuries. Three centuries is a long time in the annals of America. It is time enough to build a city—time enough to marry bricks and mortar, cornice and cobblestone, roof tile and chimney pot; twisted streets and squares and avenues; fences, bridges, and trees; Common and parks; docks and markets and graveyards; color and flavor and vista; spires and the sprinkle of bells; time enough to contrive a city by the sea and give her true grace and noble definition. The blending of all these things material and immaterial was completed in an earlier century than ours, but the marvel of Boston is that the blend has somehow survived. Roundabout and even within the old town has grown the new; but the original profile is still there, like the fine Roman nose in a Copley portrait, and all the traffic and bewilderment of modern life have not changed it very much.

New York rises from the sea as the Olympic Mountains appear to rise out of Puget Sound—majestic, radiant, aspiring. Boston struggles out of the sea like a tired swimmer, and leans back comfortably against her little hills to have another look to the clearing east and the unaccountable islands in her busy and intricate harbor. Visitors to Boston find fragments of other cities in her, such as the London of Queen Anne, the murky shadow of Dublin, a little of the little old New York now all but vanished, a twist of Baltimore and Philadelphia, and the suggestion (mostly harborside) of San Francisco. Some have even detected in the Fenway region a fleeting likeness to the Paris of the Étoile. It is all quite possible. In the case of London this indicates inheritance; but the relation to San Francisco is largely a matter of contours and approach. And of course there exist certain intramural resemblances as we look across New England: corners and doorways and windows of Providence and Portland, for example. But in the end, when these compliments have all been paid, the heart of Boston remains very much an individual creation. You have not really seen this city anywhere before.

Oddly enough, the old-worldly, the sleepy and engaging view of Boston is eastward from the most recently developed angle of the Cambridge shore along the Basin—the view from Technology bridgehead looking toward the Hill. It is here in the late fall afternoon, when

> crimson-blank the windows flare

that the color and shape of Boston—the dusty dull brick red, the long respectable façade of the living-room side of Beacon Street, the myriad cluster of shelving roof and chimney ascending to the Capitol dome, the shadowy pillar of the Custom House, and the soft grey silhouette of that one beautiful bridge —give you the surprising aquatint that you never forget. Other Boston views are exciting, but this is the one for the stranger.

I describe the city in terms of the old because I want to show her as one would show a fine old canvas from which the smoke and oil-deposit and the soilure of a viscous atmosphere have been removed. This is not to say that the present is simply a veneer —a blinding layer of chromium and neon lights and modern falsefront common the country over. Nor that you have to scratch the surface from Dock Square and South Boston to Pinckney Street and Beacon in order to discover the remnants which were lately the flowering of man's ambition to build uniformly well and with an ideal behind his blue-prints. I mean to say simply that Boston is what she is today because the past is physically as well as traditionally a part of her modern life. Probably not more than two or three other cities in America still have a central area where radios, vacuum cleaners, and electric fans are supplied—as in the early age of Edison—with direct rather than alternating current. This is about the way that history flows through the wires of her vivid present.

2. It has never occurred to me to think of an Eskimo apart from his igloo and kayak. He seems joined to them, or they to him, and the picture has a

pleasant if chilly unity. So in a warmer way with the Bostonian. It is utterly impossible to think of him apart from his city. "The bluebird," said Thoreau, who lived only within the lengthening shadow of Boston—"the bluebird carries the sky on his back." Anyone with an ounce of perception will see the sharpness and beauty of that—just as Emerson saw it and quoted it in a very human essay on his younger Concord friend. Well, the Boston imprint is likewise on your true Bostonian. It is not a flake of color or anything tangible: but you know somehow that the city of his fathers has touched him, and he carries the mark of it not only on his back but in his face and bearing, in his look and speech, in his daily life. On the streets of Seattle or London, on the deck of a steamer, in the seat of a plane, on the beaches of the South, in the cold lakes and rivers far down East, you will recognize him at once by something he does or says or does not do or say, observes or ignores, and straightway you will think of the city whose signature he bears.

We have ways of recognizing genial Texans just by glancing at them. But when we have pinned them down, it is not after all to a single city but to 262,000 square miles of marvelous— well, let the Texan describe it for you as he can and will. When we pin down our Bostonian it is part of the legend that we pin him to the Hill, the Back Bay, and the State Street area. Of course this is neither fair nor correct, but what legend ever was? There is but one Boston to the world at large: the Boston of the minority of the late George Apley, now that the world has a name for him too. Long before Mr. Apley was in a position to be late, however, the controlling characteristics of his city were established and clearly evident, and the typical citizen had already become the chain-joker's delight. Story led to fabulous story; but mark you, as with those aimed at the dour and craggy people of Scotland, there has always been the suspicion of respect in the telling of them.

Now why is this lean, conservative, shy, rather silent, well-educated, well-preserved, and apparently provincial person, whose accent turns on the letters *a* and *r*, the man he is? I venture it is because he is conditioned from childhood to fit his city as the glove fits the hand. He is, among us all, one of a really happy breed in a happy environment. Odd things are said and thought about him. Apartment dwellers of other and newer cities find it hard to believe that he often lives in the house in which he was born or raised, or in a house in which other Bostonians known to him were born or raised. And if he himself lives in an apartment because he is young, or now that the children are married, or now that the world has changed, he has a shrewd and fairly accurate notion of the history of the plot of ground on which his building stands. A British friend of mine who is very much of a seasoned journalist remarked on first seeing New York City that it was inspiring but temporary in appearance. He did not say that of Boston.

It is probable that the simple and natural act of inheriting possessions—whether land or a business or a share in the Athenaeum, or portraits and Colonial silver and Willard clocks—tends to make one conservative. All over America today youngsters are commencing married life in the twentieth-century way with spineless sofas, legless chairs, and a cheap reproduction of a chrome-yellowish van Gogh in a driftwood frame. None of these articles, however comforting, can whisper, "I was buried under an elm tree during the Revolution." But that kind of whispering is heard by more people than a shout from the housetops.

And then this matter of provinciality. The maritime history of Boston, like the very names of India Wharf and Atlantic Avenue, and the clipper shaft to Donald McKay, suggests the cloudy ships of another century when world landfalls and departures were in one way or another a part of his daily affairs. Has not the Bostonian of today well earned his insularity, if we may call it that? He has been abroad and got over it. Why laugh if he says he travels to Europe by way of Boston Light and to California by way of Dedham? If you think him ignorant of life and manners west of the Hudson, let alone west of the Mississippi, remember that Francis Parkman was out there some time before you and made some notes about it in *The Oregon Trail.* If he summers, as is likely, within sight of the beam from Minots or Thatchers, let it be said that his recent pioneering brings him drinking water from twice that distance to the west. And if he appears at times a little too conservative, remember that his city is full of charitable and public institutions and museums to which he is generous and decently anonymous in his giving. If he is one of the nation's great bankers, he is also one of its great doctors and educators and lawyers, and he has made his name in letters and in art. I need not add that he is one of its great historians, for is he not history itself?

CLEAR AND HAZY As we look at Boston and Bostonians, we should at this point look at Boston weather. It is impossible to overlook it. In the Aleutians weather is a menace, in Los Angeles a monotony, in England a mistake; but in Boston it is simply a problem. Old Bostonians never mention it; but not so the stranger. He speaks of it in comparative terms. It is wetter than Kansas, drier than Washington, colder than Virginia, hotter than Minnesota, clearer than the Labrador, foggier than New Mexico, and stranger than all get out.

Farmers, woodsmen, plainsmen, and mariners are all instinctively conscious of wind direction and can read the sky. Your Bostonian, almost alone among city dwellers, can read the sky even when he can't see it, and he knows the east wind on two good counts—by its cooling quality and the faint and, to him, quite pleasing smell of distant fish. Boston is the city that the hurricanes curve out from, thunderstorms come up in, the snow comes down on, and in which the ice unfortunately likes to linger. In the dead of winter the cars parked on either side of the narrow, devious streets stand a little farther from the curbs each day as the local glaciers force them out, and you can measure the intensity of the weather by the speed at which you can drive between them. The snow, we like to think, is rapidly removed. I wish it were. I wish this view of mine were true:

> In Boston when it snows at night
> They clean it up by candle light.
> In Cambridge, quite the other way,
> It snows and there they leave it lay.

The early Bostonians, like all early New Englanders and rough settlers in general, took a realistic view of the weather—particularly winter weather. They had to. They were not riding to and from their oil-heated destinations in taxicabs with a two-way radio, or by subway or by bus. They were out in the weather, and they well knew that the weather was out for them. Yet they had a picturesque notion of describing it in their diaries and journals and letters. There was no grumbling. The face of Boreas at the window was, after all, much better than the face of an Indian. I have been reading in some of these diaries and journals, and here are a few entries which show a poetic turn of phrase not without an overtone of cheerful optimism. In the admirable *Diary of Samuel Sewall* for February 1, 1699, for example, we find him saying:

> A pretty deal of Thunder, Rain, and Hail the last night.

April 12, 1712, on the other hand, was a much brighter day for him as he happily observes:

> I saw six Swallows together flying and chipering very rapturously.

And then John Winthrop, the Colonial scientist who faithfully made in the neatest of hands the most careful meteorological observations at Cambridge, is even more adroit in his choice of words. Here are five random entries about Cantabrigian weather, which at times closely resembles the weather of Boston across the Charles. They were recorded between the fine old years of 1744 and 1747:

> December 18, 1744: A flight of snow in the night—now very fair.

> January 8, 1745: Pellucid hail.

> March 14, 1745: A flight of hail.

> February 23, 1746: Close thick weather.

> May 11, 1747: Covered & foggy. Soon clear & hazy.

You will grant that this is all very charming to us of the braver atomic age. Observe that it is a *pretty* deal, not a great deal, of thunder; a *flight,* not a fall, of snow; *pellucid,* not damaging, hail; *close thick weather,* not rotten weather. It all sounds quite bearable. At one point John Winthrop is not unlike us in his modern view of the local elements. "Clear and hazy" has a familiar ring, but it is hard to explain why unless you have lived in the clearness and haziness of Boston itself.

Everyone remembers that the snow was deeper in the years of childhood than in any years since.

Everyone forgets that two feet of snow was *very* deep when one was only four feet tall. Two feet of snow today is just over your galoshes if you have any.* All the same, the fearful winters of our childhood and the more fearful winters of our grandfathers' day—at least according to the way those grandfathers described it—make us wonder whether we have not entered a new and softer phase of the New England climate. The late Robert De Courcy Ward, Professor of Climatology at Harvard, asserted with some conviction that weather and climate have not changed from the time of the landing of the Pilgrims down to the present day—the present day in that case being 1922. For all the old reports of driving "sixteen miles in a large double sleigh upon the crust of snow across fields and pastures," and of finding six feet of the stuff in Boston in December 1786, when people were employed to level it in the streets—in spite of these reports, we may read in the same diaries and journals any number of descriptions of mild and open winters—of green Christmases, of wild violets in bloom, and of lilacs "throwing out their leaves" in January. In a winter some three hundred years behind us, the Reverend John Eliot, the Apostle to the Indians, reported that there was "no snow . . . nor sharp weather," so that it was possible to "go preach to the Indians all this winter, praised be the Lord."

So much for the old days. What have we now? Why, we have the loveliest springs in North America. They come with a rush and take your

*Three feet of snow covers Boston as I edit these paragraphs in the month of February 1948. Three feet on the level, that would be, were Boston on the level; six feet at the crossroads—if you can find the crossroads.

breath away. We have autumns to blow the heart of Vermont and the reds of maples and the yellows of the birch right into the doorways of Tremont. And we have summers—just a little hot and sticky at times, but cooled any minute by that wonderful east wind. And we have winters—but winter is over and let's forget it.

Think about this instead. If Boston cannot claim to have the most delightful year-round climate imaginable, she can claim one of the very best climates for productive work, physical energy, and general well-being. If I read correctly some convincing statistics compiled by Dr. Ellsworth Huntington of Yale, the average optimums of climate for physical and mental efficiency suggest a yearly temperature average of about fifty-one degrees. Boston, with 49.6 degrees, is close to this—much closer than Florida, the Riviera, the summer Alps, and southern California. Other factors in the ideal climate are frequent variation in temperature, moisture, and wind; and if you don't think that Boston is close to the ideal in this as well, just look her up along the forty-second parallel.

LEVEL MEASURE Any school child in South Dakota can tell you that Boston is the home of the bean and the cod, though not many Boston school children can tell you that South Dakota in turn has its own formula for an inland herring salad. Boston has more in her refrigerator than the average American may think. Her immediate neighbors know it for a fact. She is the capital of a state bounded by Atlantic seafood and by five other states, each fiercely jealous of its own produce or its own way of cooking it. Boston is not wholly unmindful of the New Hampshire blueberry, Vermont maple syrup, New York cherry pie, Connecticut corned-beef hash, Rhode Island johnnycake, and cranberry juice from the red bogs of the Cape. With all her varied population and mixture of nationalities, one might wonder

> That the bean and the cod of the fable
> Have held up their end of the table.

Of course they have. The beans still come in individual pots just a shade lighter in color than the companionable brown bread, which ought to be toasted, but generally is not. The cod industry has so grown and flourished that fish cakes are now as broadly American as rice pudding, though most of our countrymen have not yet learned to pour a thimbleful of cream on top.

Of course when it comes to be served, it actually isn't cod but scrod. Thousands of visitors to this city are annually fooled into thinking that they have avoided the sand-white symbol of the state and depart with the happy notion that they have, on the other hand, discovered an entirely new variety of fish. It is pretty difficult to tell Virginians exactly what is meant by Lake Winnipeg smoked goldeye. It is equally useless to explain to Nebraskans the age, nature, cutting, and preparation of scrod.* You just eat it with a drop of lemon and know that it is good.

Boston is a great tripe center. With the passing of the old Adams House and Young's Hotel, two legs of the tripod have disappeared; but the Parker House still serves it in its pseudo-waffle shape, without syrup; and many other places carry it clandestinely on the menu. Of course it has a Rotarian function in Philadelphia pepperpot. Boston cream pie is a kind of institutional dessert, somewhat overfeatured in the dining halls of local schools and colleges. The Parker House roll is almost as famous as the Bunker Hill Monument, though not so easily placed geographically; and the Porterhouse steak, a name full of juice and priority, has long outlived its forgotten proprietary tavern in the city across the Charles.

Baked Indian pudding is a delicious Boston spe-

*I don't understand it myself.

cialty, particularly when mildly warm and topped with a Durgin-Park quality of vanilla ice cream. Doughnuts and pie for breakfast are still possible and even popular factors on the greater Boston bill of fare. If doughnuts at eight o'clock in the morning seem odd to the uninitiated, that opening wedge of squash or apple is actually something that the most hardened tourist can scarcely take. Which is why Boston sightseeing begins in the railroad lunchroom. Here the doughnuts are under glass, the pie is under the counter, and the man who orders simply bacon and eggs is under suspicion.

Bostonians are conscious of the enchanting quality of their fish chowder. A good Boston fish chowder is hard to beat on Monday or Tuesday, and impossible to beat or even substitute for on Friday. Boston clam chowder—to wade in shallower water—is the pure elixir from the sea. It bears no resemblance to the Manhattan transfer which involves a powerful tomato base reflecting the presence of the stiff or wire-haired variety of clam: a dish wholly unpalatable to anyone living within gunshot of Dorchester Heights, the Hill, or the Fenway. Quahog chowder is simply clam chowder seen through a magnifying glass. It is ordered, eaten, and understood largely by natives of the fifth generation.

Across the country, wherever cooking is held to be an art and not an industry, the food which Boston prepares and consumes is sometimes thought of as ruggedly simple and ingeniously unflavored. Such is far from the truth. The market of Boston is the market of the nation. It may not include the Pacific sand dab, the Gulf-borne pompano, the Oregon blue grouse, and such; but you will find there nearly everything else from the small and delicious

Malpeque oyster and Gilfeather turnip to Philadelphia scrapple, samp, and Wisconsin cheese. Boston restaurants are among the most praised eating places of the land. Durgin-Park, the Union Oyster House, Locke-Ober's, and Jacob Wirth's—to name but four, conspicuously unconnected with hotels—are as unalike as they appear to be remarkable in atmosphere and culinary delight.

And one thing more. At a time when bookstores everywhere are flooded with cookbooks of general, local, regional, foreign, offshore, barbecue, and highly specialized interest, it is still the home of the bean and the cod which sponsors the daddy of them all. Fannie Merritt Farmer's *Boston Cooking-School Cook Book* of nearly nine hundred well-seasoned pages has lately celebrated an anniversary with a golden jubilee edition. First published in 1896, and currently edited not by the Farmer's daughter but by the Farmer's niece, the house of Little, Brown (the more than century-old publishers up on Beacon Street) record a total sale through the years of well over two million and a half copies. The sales of this cookbook, in fact, have been exceeded only by such lively competitors as the Bible, *Uncle Tom's Cabin, Gone with the Wind,* and *How to Win Friends and Influence People.* Herself the "Mother of Level Measurement," Fannie Farmer gave America the teaspoon clue to creative cooking.* Her achievement, while perhaps not in a class with the discovery of atomic weights, took much of the guesswork out of one of the blessed and most aromatic of

*A persistent myth reports that in a new edition of the *Cook Book* appearing a few years ago the publishers were anxious to include recipes for cocktails. The revised copy was ready for the press when someone discovered that the formula for a Dry Martini began: "To one cup of gin add . . ."

the arts. It may be that thirteen guns were fired in Market Square in 1784 for every toast proposed at the banquet to Lafayette. But from here far out to that other ocean millions of meals every day are in themselves a toast to a lady of savory memory who was our democrat of the breakfast table.

LONG PATH If the casual visitor to Boston remembers but one engaging landmark in the city, the chances are it will be the Common. Every city has a large park of some sort, or many parks; but it or they often resemble an afterthought. As if someone had said: Now that we have a town, what about a park? Old as Boston appears to be—old in the American sense, that is—Boston Common has an even more venerable look to it: a true inner landscape from which the colors have faded just enough to satisfy the antiquarian in each of us. Certainly no other American city can boast a slice of the wilderness so tinctured with history as the fabulous area bounded by Beacon, Park, Tremont, Boylston, and Charles streets. And certainly few American cities encompass a major park so reflecting the essential country much as it was in the beginning.

Now it is hard to do business or to shop, drive, walk, or even stroll in Boston without passing or crossing the Common at some point in one's itinerary. And the Common has always something to tell in return. It compiles a fairly reliable index to the unreliable seasons. In the winter it is the last place to preserve an inch of snow the surface of which even remotely resembles suburban whiteness. It has slopes good enough for coasting, and tradition (did he know it) to support the young coaster. In summer it is full of leaf and shade. It is true that the grass tends to burn a little on the crown and down to westward. That is perhaps because the sun not only shines above it but is also reflected on it from the many walls of brick and glass. At best, it fares somewhat better than the average lawn of the person of average means. And why not? Every Sunday, when the weather is fine, thousands of men, women, and children eat, loaf, sleep, play, or seek romance and escape on every available square yard of that grass; and when they come to leave they carefully, almost reverently, cover it completely with the brighter parts of the Sunday supplements. How queer to think that in the casual old days their predecessors went to pillory on that same ground for acts of milder uncivil disobedience. About the middle of the seventeenth century, in fact, a twenty-shilling fine was imposed on citizens of the town for spreading trash upon the Common. How quaint, how footless they were!

Here at the spring, the casual commoner is more interested in green blades, green buds—in pigeons, squirrels, and peanuts—even in the Latin names of *Ulmus, Quercus,* and the like, stenciled against the trunks of the inarticulate trees, than he is in the history of this half-hallowed, half-unholy ground. Perhaps he is sitting on a bench. If so, he sits on tradition. Exactly one hundred years ago "new iron seats were placed in the Common. They prevented whittling."

It is impossible today to estimate the value of the Common to Boston. It is a value tangible enough in one way. But its spiritual component is also great. Conversely, it is staggering to think what the land would fetch were it up for private sale! The Reverend William Blaxton,* first lone settler of the Boston promontory, who used to ride about his land on the back of a bull, in 1634 became convinced that he should flee farther into the wilderness from the lord-brethren, just as he had fled from England to escape the lord-bishops. But before he fled, he sold to the town the piece of land which had been set aside for his perpetual possession, keeping only for himself some six acres, the definition of which has long since been lost in the architectural geometry to the north of Beacon Street. The parcel which he sold to the town comprised some forty-eight acres, with boundaries then about as now, and they

*Scholar who loved, and therefor left, the most,
secret and solitary, no Indian-giver,
who to his own cost played the generous host
and asked adventurers across his river:

what would he make of us, if he could see,
after so many tides have ringed this coast,
what manner of men his children's children be
to welcome home his still inquisitive ghost?

He, more than all, of individual grace,
the pilgrim innocence, self-knowledge sure,
stepped like an angel in this savage place,
and, in all nature, found no evil-doer.

A summer's freedom on a bramble shore
whose wild rose the Lords Bishop could not blight,
then the Lords Brethren saw him close his door,
bidding his orchards, and his house, good-night. . . .

—From *The Kid,* by Conrad Aiken

cost the town just thirty pounds, or one hundred and fifty dollars.

Some fifty years later four aged men deposed that, immediately after the purchase from Blaxton, the town lay out a place for a training field. The tract in general was used for the feeding of cattle. What a strange mixture of the provinces of Cybele and Mars! Town records show it was ordered that the inhabitants should keep but seventy milch kine —no dry cattle or young cattle, and but a single horse. This would be in 1646. The horse belonged to elder Oliver, but who was he? A man might keep four sheep in place of one cow. There was a keeper of the cows who received initially two shillings a cow for his work. He shared his vigils with a shepherd who got eightpence for sheep and threepence for lambs. While these cows and sheep and lambs were in self-training to survey and lay out the streets of the future Boston as we know them, the militia was drilling away one day a week. Cows and sheep and militia! A combination of interests not without mutual danger. General Humphrey Atherton, returning home on horseback from reviewing troops one evening in 1661, was undoubtedly well mounted for the journey. But he either relied too much on equine judgment under a loose rein, or at some point the curb failed to balance his spurs. At any rate, he struck a cow in the darkness on the Common and was killed.

On Easter Sunday, west of the Common along the Mall on Commonwealth Avenue—which was nothing but unsubstantial water down to the middle of the last century—you may watch or take part in the annual Easter parade. A far cry from that earlier century, when even on the hottest Sundays in midsummer the people were forbidden to stroll on the Common! The cows were still there, but if any horse was observed on that hillside on the Sabbath, the owner was obliged to part with the sum of five round shillings.

What other green pastures ever equaled these? And what other plot of American ground across three centuries has known so much of freedom, persecution, dreams and politics, poetry and war, wretchedness and exaltation?

2. Crossing Boston Common in the new warmth of an April day, the world seems suddenly as innocent as the blue-gray flocks of pigeons in our path, and as gentle as the light breeze ruffling the newspaper of that old man asleep on a bench. Innocent, gentle, and warm: and the mind gone hunting on the breeze. For how few of us who make this cheerful excursion are aware of much, if any, of the history of the ground on which we tread! Some of that history is pretty grim, some of it is comic, some of it is downright stirring and beyond imagination. The Long Path, so romantic to Dr. Holmes's Autocrat, was often the short path to oblivion. But Boston Common is an ancient battleground of ideas and beliefs and freedoms—unmarked by pyramids of cannon balls, unrelieved by crosses to the fallen, as void of redoubt as of the mischievous doubt in the worst ancestral skull.

The sixteen-hundreds witnessed on this spot a good deal of miserable injustice, cruelty, and the awful shadow of the noose. The hanging of four Quakers simply for being Quakers is perhaps the darkest chapter. Of the three men most certainly hung on the Common, we are likely to forget the names; but the name of Mary Dyer is with us still. She was reprieved at the last possible moment, packed off to Rhode Island, only to return to Boston fanatically to die. Tradition, if not all the facts, suggest that she, too, died on the Common. This was on June 1, 1660. The hangings on the Common that began in the sixteen-thirties—relieved by executions in the military manner, though not all for desertion and military reasons—were depressingly frequent for a long time, tapering off to zero about the year 1812.

Digging sods was forbidden on the Common, but staining the sod was not. The American King Philip visited the Common in 1662, but thirty less fortunate Indians were hung there in 1676, and nine others were shot in 1678. Captain James Hawkins and seven pirates mounted the scaffold ladder a few years later, to be followed by other pirates, and so alas! it went. Burglary and robbery offered excellent grounds for death; and along in the late eighteenth century a woman was hung on the Common simply for snatching a bonnet worth seventy-five cents from another woman, and running away. There was also the whipping post, the pillory, and dueling. Yet we may read as early as 1663 that "the Common was the beauty and pride of the Town, ever suggesting the lighter side of life." The lightest possible side would seem to have been when one's two feet were off the ground.

From the very beginning the trumpet blast has sounded over the Common, and queer and colorful uniforms seem always to have added their animated accent. La Tour and a company of French soldiery were there in 1643. The three thousand soldiers who embarked from Boston with Sir William Pepperell in 1745, to proceed against Louisburg, camped on the Common. From 1768 to the evacuation of Boston by the British eight years later, the redcoats were almost continually encamped on the Common, with a regular garrison of 1,750 in the winter of 'seventy-five and -six. Men who were boys in the first quarter of the nineteenth century played round the as yet unleveled lower fortifications and the redoubts on Flagstaff Hill—where the Martin Milmore Soldiers and Sailors Monument stands today. In large part, the British troops who fought at Bunker Hill faced off into history from the Common. In 1824 Lafayette was there, firing a cannon and hitting the target floating on the water over what was later to become Berkeley or Clarendon Street. In 1860 King Edward VII, as the young Prince of Wales, reviewed our troops in the northwest corner of the Common. It is interesting to reflect that on that day, in the uniform of a colonel, he rode a handsome horse named Black Prince which now, in heroic green bronze likeness, carries General Washington through the Public Garden on his endless survey of the west.

In the Civil War it was the same; and the First Massachusetts departed from the Common in June 1861 and was mustered out on the same ground in 'sixty-four. All through that saddening lustrum, the Common was the scene of recruiting, departures, heartbreak and weeping, and thirty-five- and one-hundred-gun salutes for battles won. Now, in our own century, the First and Second World Wars have had their reflection in the Common. Part of the iron fence is gone—which opens the view. At the very moment of writing these words there is an

Army mechanized display just east of Charles Street, its searchlights figuring the sky; and the Bay State Club, which seems completely and permanently at home on this once martial ground, reported an astonishing attendance count of 29,000 members of the armed services for the first quarter of 1947.

Looking back on the pastoral side of it, the little Frog Pond with its modern solid bottom is worth a glance. How pleasant to imagine it, under the old willow and full of frogs and shiners—even after it was "curbed and improved." Some boys—perhaps the first—were fined for bathing in it a hundred years ago. Somewhat earlier a young boy named Francis Parkman commenced his nature studies on the Common "by sounding the mysteries of the Frog Pond with a pin hook where he and his cousin caught horned pouts which they seasoned with ground cinnamon and broiled over a fire" in the Parkman garden.* Other boys playing ball near by unearthed a red fox from his hole. Ninety-nine years ago, the great Water Celebration, when people laughed, cheered, and wept, marked the introduction of Lake Cochituate water for municipal use. When the valve was turned, the soaring column in the pond where Parkman had fished rivaled the lofty plume rising from the beautiful lake at Geneva in Switzerland. The place was ready for big things. In 1859 an elephant, belonging to one Sam Rice, bathed in the Frog Pond.

It is impossible even to attempt a three-hundred-year summary of the uses to which the Common

*An excellent preparation for his Indian days to come not so many years later. From Chapter XIX of *The Oregon Trail:* "A small boiled puppy was served up as a parting feast, to which was added, by way of dessert, a wooden bowl of gooseberries from the mountains."

at large has been put. Emerson as a boy drove a cow to pasture there; and cows were not put on the Common black list until 1830. Fireworks, kite flying, snipe shooting, carpet beating, Punch and Judy shows, telescopes, coasting the long coast on what were called monster double rippers—one of them worth two hundred and fifty dollars—the Common has known them all. Whig mass meetings, antislavery meetings, a railroad jubilee, promenade concerts, conventions, Indian dances, sermons, the secluded smokers' circle, three hundred spinsters spinning, a wishing stone, ropewalks, Sunday orators, ox roasting, horse racing, hoop rolling, balloon ascensions, a place to spread infected clothes during the smallpox epidemic, a refuge from the great fire of 1872—the Common has seen, known, heard, or been them all. Subways, buildings, and other improvements have now and then threatened to alter the familiar landscape. The cry of "Save the Common" has always prevailed. It will probably prevail against the underground movement for a subterranean municipal garage. I trust so.

BRONZE Some years ago one of the most thoughtful and respected writers in England, who generously acknowledges his principal literary debt to Herman Melville and Thoreau, was visiting Boston. One morning he happened to pass a reasonably odd statue not far from Ashburton Place. A figure queer enough for anyone, it held Mr. H. M. Tomlinson's attention for a silent minute or two, as though it were spastic or epicene. I supposed he had seen worse at home. Apparently not. "Rum old bird," he said, and walked on.

Boston is a city full of statues, not exactly in the gay and ornamental and even triumphant sense that Paris is, but more in the severe and frugal manner as in London. It is not disrespectful to say that a lot of rum old birds look down at you from their rum old pedestals. Wendell Phillips in his day could and did make some harsh remarks about Boston sculpture. But on the whole one can find among our share of bronze and marble a surprising number of good examples of American civic art—most of it outdoors, some of it within four walls. We all have our favorites. It is too much to think that mine will be yours, but the few that I shall name mean something to me quite beyond the boundary of the medium in which they are cast. A statue does not necessarily have to conform to the highest standards of art, somehow, to be compelling. It ought to, but it doesn't. The subject, the setting, the size, the weathering, and the quality of the patina are all in themselves factors of interest. A fine old piece of bronze gone green as the lichen on a forest rock or tree is an object of art in itself. The signs of the zodiac on the floor of the Public Library just inside the front door are lovely enough for the yellow shine and graceful curve of the footworn metal.

At least three statues of varied and popular interest are a part of the Public Garden. The Ball equestrian statue of Washington is generally acknowledged to be one of the fine examples of its kind in the world. It is not as vivid as Anna Hyatt Huntington's Joan of Arc in Gloucester and elsewhere, for example, but it has about it the sweep and grandeur of the heroic. It has both life and dignity without a suspicion of dullness. There was an earlier project for a Boston equestrian statue of Washington, but Thomas Crawford, who was to do it, died in 1857, two years before the Public Garden as such came into being. The commission accordingly went to Ball, and he used as a studio a barn erected on the Tremont Street factory site of Messrs. Chickering & Sons, the piano makers. It is curious that the success of most equestrian statues turns largely on the horse. The horse is the one animal in art that remains perennially alive. Up at the State House, when you look next at the excellent figure of Hooker and his charger, remember that you are looking at the admirable result of wise collaboration: Edward C. Potter modeled the horse, and Daniel Chester French the figure of the general. At the other end of the Public Garden, on much too small a pedestal, stands the familiar figure of Edward Everett Hale, hat and walking stick in hand. Bela Pratt was the sculptor. I do not admire it as a work of art—nor do I admire the sculptor's figures facing Copley Square—but there is something homely and simple about it which I do admire, and it is my friendly guide what days in summer I wander by to inspect the leafy umbrella and the clean, spare branches of *Fagus sylvatica,* a well-remembered beech tree only a few rods to the north. My third figure in this area is the splendid Kitson likeness of Kosciusko—a forceful and living creation which reminds one of the romantic qualities in the more heroic interpretation of Sir Henry Vane—"that true friend of New England"—signed by MacMonnies and all but hidden away in the Public Library.

Everybody who knows Boston knows the great Robert Gould Shaw memorial, standing where the northeast corner of the Common faces the State House. Full of motion and kinetic power, it is almost equally a memorial to the genius of St. Gaudens, the sculptor. Down the Commonwealth Mall a series of great men face into the east. Alexander Hamilton by Rimmer, subdued to stone but strong in outline, is the first American that Washington sees ahead of him from his lofty saddle inside the Garden gate. Beyond Hamilton he would ride past an old supporter in Martin Milmore's General Glover; and beyond Glover, the vigorous Garrison who said, "My country is the world, my countrymen are all mankind." Far over in the Fenway delta, near the Massachusetts Historical Society, is the John Boyle O'Reilly memorial by Daniel Chester French, its effectiveness greatly enhanced by the screen of Celtic weave against which it has been so attractively set. Cyrus Dallin's Indian dominates the main entrance to the Museum of Fine Arts, his Paul Revere charges handsomely over the crown of the North End where he lived but did not ride, and a capable Farragut by Henry H. Kitson looks seaward as he should from the vantage point of Marine Park in South Boston.

I am one of those odd people who investigate all statues within range for whatever light they throw on the subject or the sculptor. Down on the Esplanade, inspection of the charming memorial fountain to Lotta Crabtree shows what a skillful artist can do in the animal kingdom with the simplification of line and plane. This fountain is the comparatively recent creation of the Bostonian, Katharine W. Lane, whose much more important work as a corporate part of Harvard's biological laboratory is worth a special trip to Cambridge.

The day of the frequent unveiling of outdoor statues appears to have passed. Mindful of all that we have in the wide field of this most durable of the plastic arts, I often wonder why the city has not commissioned, or no artist has attempted, some rousing tribute to Parkman's thundering history of the Oregon Trail,* some authentic reflection of Thoreau in his cosmic loneliness, some durable suggestion of the golden age of sail. We have not yet outrun our past, and I cannot believe that we have run out of the dynamic in creative, three-dimensional art.

*I mean beyond the attractive, simple Indian monument out at Jamaica Pond. I have mentioned the existing shaft to Donald McKay, but the spirit of the clipper ship remains to be caught—and caught on the clipper-ship scale.

INTRAMURAL If New England is the center of higher education in the United States, then Massachusetts in general and Boston in particular are the focal point of that center. Some seventy-five institutions in this state are authorized to confer degrees—a larger number than you will discover in any other state in the Union. More than half of these seventy-five lie within the area of metropolitan Boston. Furthermore, Massachusetts stands at the top among the forty-eight states in the total value of endowment funds in the possession of its colleges and universities; and it is safe to say that the preponderance of this amount is centered in Greater Boston. I shall not identify them all, but the local list is made sufficiently impressive simply by reciting the names of those that come easily to mind: Boston University, Boston College, Northeastern, Harvard, the Massachusetts Institute of Technology, Tufts, Simmons, Jackson, Radcliffe, Wellesley, Emmanuel, Regis, Emerson, Calvin Coolidge, and Suffolk and Portia Law Schools. Beyond these are the legion of institutions specializing in the arts or sciences, among them the New England Conservatory of Music, School of the Museum of Fine Arts, Massachusetts School of Art, the Fletcher School of Law and Diplomacy, Forsythe Dental Clinic, Babson Institute, Massachusetts College of Pharmacy, and so on.

The history of education in Massachusetts goes back, of course, to the founding of the Boston Latin School in 1635—the first free public school in the Colonies. The following year saw the founding in Newtowne—afterward called Cambridge—of the first college in the New World, named in 1639 for John Harvard, the young Charlestown graduate of Emmanuel in that other Cambridge across the Atlantic. It was he who left the new-born institution half his property and all his library. The more than three centuries which have passed since the opening of Harvard College have been starred by the birth of so many sister and neighbor institutions that a casual tour today of Greater Boston is in part a kaleidoscopic view of towers, spires, and domes; recitation halls, dormitories, sequestered grounds; fairly ancient elms and not so ancient ivy, and athletic fields to match. Metropolitan Boston has offered the setting for all this, but in return our many institutions of higher learning have, on the material side, enlivened and improved the landscape, and on the intellectual level have enriched the community and insured the continuing renewal of certain spiritual and humane values vastly to its credit.

Most intimate with Boston proper in domicile and strategic points of contact is the highly complex structure of Boston University. From that day in Boston in 1839 when some officers, graduates, and friends of Wesleyan University down in Middletown, Connecticut, met and founded a theological school which was the nucleus of the Boston University to be, this institution—for the past twenty-two years under the presidency of Daniel L. Marsh —has grown and multiplied to an extent of which few Bostonians are fully aware. Out along Commonwealth Avenue on the riverside, between the present limits of Granby Street on the east and Ashby on the west, the small pointed tower of the handsome new modern Gothic building on the campus (over which were fired some guns of the Revolution) marks the heart of the Boston University of today. As Dr. Marsh has pointed out, until the development of the Charles-side area—which began but a few years back —the University, in its myriad departments, had lived the life of the hermit crab which finds its home in the empty shell of some other marine creature, moving into new and larger shells as growth demands. With the exception of what is called Robinson Chapel, the departments of the University have always occupied buildings previously used for other purposes.

Gathering unto herself a number of departments which had sprung to life subsequent to 1839, Boston University as such was formally chartered in 1869. As much an organism as an organization, this fluid state of a growing institution has given it certain marked advantages. Boston University has never lacked the will to experiment or adapt herself to changing conditions, either physical or curricular. She early elected to make her home not in the country or in a suburb but "in the dark gray town," as Longfellow said. She has touched and still touches many Boston streets. Take the original School of Theology. First established in Newbury, Vermont, it was soon moved to Concord, New Hampshire, and then to Boston, where it found shelter successively on Pinckney Street, Bromfield, and lastly on Mt. Vernon, just where you will find it today. The University's College of Business Administration saw the light in the old Massachusetts Institute of Technology building on Boylston whence it was removed in 1939 to the Charles Hayden Memorial Building on the new Commonwealth Avenue campus. The College of Music, established in 1872, went the rounds of five locations. The Law School has known Bromfield, Beacon, Ashburton, and Ashburton Place, its present home; and the School of Education has flitted from two different numbers on Boylston Street to the Mechanics Building, to the old Massachusetts Art School Building, and finally to the Soden Building at Exeter and Blagden streets.

Boston University is about one third the age of Harvard, but it is still a far cry back to her own founding fathers—three businessmen whose names were Lee Claflin, Isaac Rich, and Jacob Sleeper. I like to think especially of Isaac Rich, son of a Wellfleet fisherman, a youngster who came up to Boston from the Cape to sell fresh cod—allegedly the first man to blow a fish horn in her streets. I like to remember that Alexander Graham Bell, as a teacher at B.U., carried on all the experiments which led to the invention of the telephone. I like to reflect that B.U. was the first completely organized university in the world to admit women to all departments on equal terms with the men. She was the first to initiate certain other interesting ideas in the teaching of law, medicine, and theology. At present, with some 30,000 students, strict standards of scholarship, and a forward look in her eyes, she proudly boasts that, though founded by men of Methodistic faith, she has always been and still is broadly tolerant of all religious groups. Her normal enrollment includes more Congregationalists than one will find at Amherst and Dartmouth, both of them established by Congregationalists; more Baptists than in the Baptist colleges of Bates and Colby; more Episcopalians than at Trinity; more Methodists than in the two American Wesleyans combined; more Unitarians than at Harvard, and more Roman Catholics than in Boston College.

A tolerant city, presenting at every point in the metropolitan compass groups and individuals of differing faiths, the University which bears the name is typical evidence of that inner and inmost coöperation between freeborn men and women, without which we would not be very close to democracy, and with which—and only with which—we may move forward without fear and with resolute hope.

MUSEUM PIECE Mention of the word museum is for many people a reminder of what the late Robert Benchley used to call "museum feet." In this flexible age of science no one as yet has perfected a method whereby the wanderer in museums can emerge with feet still able to function as painlessly as when he entered. Achillean instep is one reason why man goes to a ball game or plays golf on elastic fairways or drives his car out into the country on Sunday afternoon instead of inspecting Degas, George Innes, or Colonial masters. Yet some progress has been made. An ingenious architect out in Omaha (the late Alan McDonald) designed the handsome museum in that city with the human foot in mind. The floor of each room has a different surface from that of adjoining rooms, which enables the visitor to exercise a fresh set of foot muscles as he advances from the primitives to Picasso. In one room the surface is hardwood, in another square tiles, in another something springy and synthetic that only a chemist would recognize, and so on. I have walked over these floors with a minimum of fatigue. When Walt Whitman suggested that we loaf and invite our souls, I sometimes wonder, etc.

Now Boston is a museum city, and within the circumference of what we are pleased to call Greater Boston, you can almost name your subject and find a corresponding museum from geology and archaeology right down the primrose centuries to the luxury of modern art. Collections of almost anything, that is, from Indian artifacts, Chinese bronzes, glass flowers, Egyptian tapestries, and relics of the age of Moby Dick, to fluorescent minerals and electronic tubes.

But the jewel of all museums in Boston is, of course, the Museum of Fine Arts on Huntington Avenue. Founded by public subscription in 1871, the gift of a piece of land on Copley Square from public-minded citizens, and the surplus of a fund subscribed for the erection of a bronze statue of Edward Everett, it offers, in its present building, one of the great collections of art in the world today. "In addition to its many galleries of paintings, and Early American rooms," says George H. Edgell, the director, the Museum "has noteworthy permanent exhibitions of Egyptian, Greek, and Roman art. Its Asiatic collections are the finest in the Occident. There are French and English period rooms, Gothic and Renaissance art, American Colonial and contemporary painting, silver by Paul Revere and other early craftsmen, a famous series of portraits of American patriots, from Washington to Samuel Adams. The Print Department is the largest in the Western Hemisphere." The present imposing building is, in fact, not a single museum but a group of several, each devoted to collections of one origin or of one character, and each accessible without traversing any other. Halls and loggias and open courts connect these various departments, and the visitor moves about (as one should in a museum) conscious not only of the art objects which he has come to enjoy but of surroundings and areas of natural beauty against a background of architectural dignity.

On March 20, 1947, the Museum rightly paid tribute to the Boston Athenaeum, which antedates the Museum of Fine Arts by more than sixty years and which was, in point of history, Boston's first art gallery. For four years prior to 1876, the collections of the Museum, both gifts and loans, were exhibited in two rooms of the Athenaeum. In that year the Museum opened its first building in Copley Square, where it continued until 1909, when the Huntington Avenue building was ready for the public.

About half a million people visit the Museum of Fine Arts in Boston each year. You would probably find, on looking up the figures, that visitors to the National Gallery in Washington in the same period exceed the two-million mark; that visitors to the Metropolitan Museum in New York exceed one and a half million; and that more than a million persons cross the threshold of the Art Institute of Chicago. Now the Museum of Fine Arts in Boston is entirely supported by private gifts, bequests, and annual subscriptions. It receives nothing from the city or the state. Compared with the more than 15,000 people who subscribe annually to the Art Institute of Chicago, the two thousand or so subscribers to our own Museum seem pathetically small in number. Here we have one of the great treasure centers of the world, open for the citizen's instruction, delight, and cultural education, absolutely free, every day of the week except Monday. Nothing in this democratic age could better represent the work of the people for the people. And yet where are the people? I have made light of the problem of museum fatigue; yet I submit that most individuals visit a museum with the same tactical approach that they use in visiting a flower show or a battleship. They must see everything at once and as rapidly as possible. The right way to go is to visit one area at a time—and one only. If you are interested in prints and drawings, don't stray into the American wing. But if you are attracted by all that is best in Colonial American art, don't spend half your afternoon among the Pharaohs of Ancient Egypt. I know many people who go to a museum to see two or three pictures or two or three art objects, and no more. I have known people who will travel to a strange city just to look at a single picture—let's say a Grant Wood or a George Bellows—owned and exhibited by the local museum. It would be interesting to know how many of the half million visitors to our museum each year are outsiders. A fairly large percentage, I should guess. One of them, at any rate, a remarkable gentleman born ninety years ago and now a resident of New York City, still devotes part of his annual visit to Boston to a pilgrimage to the shrine of Kuan-yin. Do you know Kuan-yin of the twelfth century? At least you know where to look.

CONCERTMASTER Music must always be part of the Boston legend, not because Boston has given a stream of great composers to the world, but because she is and has long been a concert and teaching center in the cosmopolitan as well as the national sense. It would be hard to analyze the reason for this, for native Bostonians by natural endowment are not more musical than people in other American cities. If anything, they are less so. But music, on the other hand, is one of the great components of the cultural pattern; and its appreciative growth, we may suppose, under the influence of a rich European culture was as natural to Boston as the growth of her nineteenth-century school of letters, and her earlier reputation in painting which flowered first in the second half of the eighteenth century in the particular genius of John Singleton Copley.

Once many years ago in Portree on the Isle of Skye in the inner Hebrides I awoke one night to hear, under the window of my dingy room in the local inn, the unforgettable harmony of a Gaelic tune whistled by three young men of the town as they walked down toward the sea. That was native music. It wasn't a wretched snatch of the June-moon school of crooners. It was something alive and deep and lovely and significant, symbolic of the poet and the artist common to a common people. I should like to say that Boston, or any American city, is related to music in that elemental way. I cannot say it, and no one can. We in America are not yet old enough as a people actually to have begun to produce a great music of our own. But we have begun to want to produce it.

But I can and do say that there is a natural love of good music in Boston that is both widespread and well-nourished. It was not born overnight. The Symphony Orchestra under Dr. Serge Koussevitzky has reached perhaps the zenith of our musical expression in a spectacular and colorful career that runs back to its founding in 1881. The Pops concerts, which are to the Boston Symphony as the better popular-priced books on the newsstands today are to their original editions, have given pleasure and delight to thousands since their beginning back in 1885. It was only natural that the Pops in turn were followed and paralleled by another and vastly more influential series in the annual Esplanade concerts in the Hatch Memorial Shell. Here under the open sky, music has the advantage of dramatic setting and the reward of audiences which have numbered as high as 35,000 people. Where the Pops concerts are doubly pleasurable for their informality and the obbligato and counterpoint of food and drink and talk, the night music on the Esplanade is listened to with rapt and undivided attention. And the heartbeat and breathing of the shadowy city add a curious note of reality as in some sonorous tone poem conscious of urban life and living.

As one historian has written: "Music scarcely had a voice in Boston until William Billings, a tanner by trade—who had been born in our town in 1746—made rude attempts to put into harmony the songs which he heard in his own soul." Blind in one eye and deformed in person, his earliest compositions were chalked upon sides of leather in the shop where he worked. His music had almost always a touch of patriotism about it. He was a choir singer, and one of his supporters was Governor Samuel Adams. It was the day of psalm-singing, and music otherwise found the town of Boston as indifferent as it then was to drama. But when Billings died in 1800, committed undoubtedly to one of the unmarked graves on Boston Common, he had written and published enough, including his *New England Psalm Singer,* to have had it since said of him that he "broke the ice which was congealing New England's music—for which America owes him a great debt of gratitude in spite of his few thousand errors in harmony." Contemporary with Billings was one Oliver Holden, author of the famous American hymn called "Coronation." It was composed for the dedication in May 1801 of the church which stood in front of his house on Pearl Street in Charlestown.

There for the first time were sung the words which millions know, beginning:

> All hail the power of Jesus' name!
> Let angels prostrate fall;
> Bring forth the royal diadem,
> And crown him Lord of all.

Holden's first book of music, *The American Harmony,* was published in 1793, when he was twenty-eight. Perhaps the composer's chief claim to contemporary distinction, however, derived from his words and music for the hymn sung by the Independent Musical Society, of which he was incidentally the director, when General Washington once visited in Boston during his administration. The hymn was sung by this sizable chorus from the top of a triumphal arch, President Washington standing on the balcony built out from the Old State House. I do not know the music, but the words as poetry are not much to remember. This hymn or ode was performed a second time at the World's Columbian Exposition in Chicago in 1893.

John Howard Payne, the actor, in one respect belongs in this picked company. Son of a Boston schoolmaster, he was born in New York City and died abroad in Tunis. But his "Home, Sweet Home," which made a fortune for his publisher though not for him, is the creative work of a man whose name is definitely linked with Boston, where he spent his childhood. His popular song is still known to everyone.

The formation of the Handel and Haydn Society in 1815 gives to the Park Street Church of Puritan tradition credit for the first important forward step in the progress of music in Boston. The Church had a choir of some fifty voices and from them were drawn the voices for the new Society. Their first public performance, however, was given in King's Chapel. Not much later, and most popular in the eighteen-forties, came a series of secular orchestral concerts provided by the Boston Academy of Music's and the Musical Fund Society's orchestras.

Italian opera loomed fitfully on the horizon. In 1850 Jenny Lind came to town—oddly enough, as we view it today, under the auspices of P. T. Barnum—to sing, more oddly still, in the Boston station of the Fitchburg Railroad. There she gave two concerts, and one thousand more people than could be accommodated bought tickets. Perhaps there was something symbolic in this, for from that time on transportation has played an increasing role, not only in bringing to Boston the best among international artists, opera companies, choirs, and orchestras, but in dispatching our own musicians as emissaries to the world.

2. There has been attributed to those two overflow concerts which Jenny Lind gave in the Boston station of the Fitchburg Railroad in 1850 at least some of the inspiration which resulted in the building of the Music Hall, first opened in 1852. In 1863, in that same building on Hamilton Place which is now the home of Loew's Orpheum, "the great pipe organ" was dedicated. Mr. Mark A. DeWolfe Howe, who knows as much about Boston history as any man alive, has said in print that by this time, indeed, there *was* a musical Boston.

In 1867 the New England Conservatory of Music was founded. Offering instruction in every branch of the art, distinguished by an able faculty and an impressive list of graduates, in the eighty years which have since passed the teaching and musical standards of the Conservatory have widely and beneficially influenced the nation. For ten years more than a century, too, the Harvard Musical Association, a club whose membership embraces the graduates of many different colleges and universities, has contributed through its concerts and excellent library to the development of music in Boston. One of its founders—and president for twenty years from 1873 until his death in 1893—John Sullivan Dwight was a powerful influence in the Association. Educated for the ministry, he was one of the transcendentalists at the Brook Farm Colony and editor and owner of *Dwight's Journal of Music.*

Opera was by no means new to Boston in 1909; but in that year, out on Huntington Avenue, the Boston Opera House was built as a home for the Boston Opera Company, an organization founded on the lines of the Metropolitan in New York. The Metropolitan had come into being in 1883 and was itself built as a rival to the Academy of Music. Prominent citizens who backed the Boston venture envisioned the new opera center as allied with the Symphony Orchestra, the Art Museum, and the Public Library in cultural value. The best available artists were to be engaged and the company was to travel to other cities. The Boston opening, on November 8, 1909, was a widely heralded event with Nordica and Louise Homer in *La Gioconda.* Henry Russell was the director. But the resident company functioned only until 1914. In the spring of that year Mr. Russell, as a final gesture, took the entire organization to Paris where it gave a two-month season at the Théâtre des Champs-Élysées. To those interested in comparative dates: the Chicago Opera Company was founded in 1910 with Mary Garden as its director, or directa. The Chicago Company and the Metropolitan, of course, have both been heard a great many times in Boston.

Of the many other remembered musical organizations, one should mention the Boston Flute Players Club, founded in 1921, with George Laurent of the Boston Symphony Orchestra as director. Its avowed purpose of furthering music for the flute, and chamber music in which the flute has a part, has been fully realized to the satisfaction of hundreds of subscribers and guests who have greatly enjoyed these concerts.

Likewise the Longy Club, under Georges Longy, famous first oboist of the Boston Symphony Orchestra, gave at the turn of the century its talented blessing to French and other chamber music. It was the Longy Club which first presented to Boston Debussy's *L'Après-midi d'un Faune*.

Today, of course, Boston's chief glory in music is the Boston Symphony Orchestra. Symphony Hall, in which it makes its home, is also the city's leading concert hall, and a single season of visiting artists offers an education in music unmatched outside of New York City. Nine conductors have made Boston Symphony history—two of them after their dynastic interval returning to lead the orchestra for additional seasons. The nine are: Henschel, Gericke, Nikisch, Paur, Muck, Max Fiedler, Rabaud, Monteux, and Koussevitzky. Under Dr. Koussevitzky the orchestra has given not only seasons of great programs of classical music, but it has introduced the city and the country, and sometimes the world, to the most interesting of the new music. In recent years, indeed, Boston has first heard through the Symphony new music by such foreign composers as Prokoviev, Honegger, Ravel, Martinu, Shostakovitch, and Benjamin Britten; and such native composers as Copland, Harris, Gershwin, Randall Thompson, E. B. Hill, Walter Piston, and Leonard Bernstein.

The orchestra has pioneered too. It was the first, under Karl Muck, to record for the phonograph. This was for Victor in 1918. It was the first, under Dr. Koussevitzky, to broadcast performances. This was in 1926 from one of our major Boston stations. Since 1942 the orchestra has given weekly broadcasts in season over the ABC Network. It was the first orchestra to organize, in connection with the now famous Berkshire Festival, a summer school of music. About one hundred and ten players are in the orchestra. The Pops concerts are given by some ninety-eight of these players—since 1930 memorably conducted by Arthur Fiedler. It is Fiedler, too, who has conducted the Esplanade concerts through every one of their appreciated twenty seasons. All this, of course, is one way of saying that Boston has pioneered in developing great audiences for music in America. None of them, however, can be an audience more enthralled than her own.

OF BLUE SALT WATER The familiar phrase of one if by land and two if by sea gives numerical emphasis to the Atlantic Ocean. So through the greater part of three centuries the emphasis of Boston has been on the sea that washes almost to her door. By reason of her long, deep-channeled, and intricate harbor, Boston is a riparian city without really enjoying in the larger sense an actual outlook on the ocean; and because of this fact we sometimes forget that she is a seaport city first of all. New York is so very nearly encircled by ships and tugs and barges and ferries, and San Francisco so plainly indented by the Pacific, that we think of them in the maritime sense first and last. Many of us in Boston today can and do go about our business without so much as a sight of any part of the waterfront for months at a time. We know that the waterfront is there. We eat and thrive on the fish and shellfish that come out of it, or the imports that arrive by way of it; but save for the deep-throated sound of a whistle now and then on a foggy or rainy day, or by courtesy of an east wind in the early and quiet hours of the morning, we might be rooted well back of the coastline—a truer Athens than we can possibly claim to be.

It was not always thus. A glance at an old map showing the original shoreline of Boston proper will explain very quickly just why the early Bostonian proudly believed himself to be, and actually was, a seaport man. Consider for a minute just where that shoreline was. The Charles River slipped quietly into the harbor, and the tide backed up into it as far as Watertown. It was fairly wide at the mouth. The line of the present Charles Street was all under water. Water came up into Cambridge Street to a point halfway between what is now the Longfellow Bridge abutment and Chambers Street. Merrimac Street and all to the north of it was water. About half of Commercial Street—the northern half—was barely on dry land. Atlantic Avenue from North Street down to Rowe's Wharf was water. Where the Custom House tower rises today there was water. South Market Street was submerged. There was water to a point just west of the site of Faneuil Hall; Post Office Square was largely tidal, and the tide ran a little west of south right up into the corner of Franklin and Federal streets. Where the South Station stands there was once no land; and where Atlantic Avenue turns into Kneeland Street, for several blocks a fish was more at home than a human being. Take away that much of the city-land that man has made, imagine the vastly more open view which one had from streets unobstructed by tall buildings, and you can readily see that it required no captain's walk to give a man a sight of the blue salt water upon which he gambled and won.

Through the more than two centuries that Boston was expanding—growing taller and wider and more congested—she was gradually shutting out the old harbor view. At the same time she was adding wharves and docks, channels, canals, and snug havens for the ships that in the 1840's were to make her unequaled among the seaports of the United States in the extent and variety of her commerce. "There was no quarter of the civilized or uncivilized globe," once wrote Mr. Frank H. Forbes, "in which the enterprise, energy, and pluck of a Boston merchant and a Boston shipmaster did not find an entrance, or from which a wealth of commerce did not return." For more than twenty years, New York, Philadelphia, and Baltimore, as well as other American business and commerce centers, leaned largely upon Boston for the produce of the distant lands below all visible horizons. Boston's monopoly of trade covered many of the leading ports in Europe, Asia, Africa, South America, the West Indies, and the West Coast. She reached the height of prosperity in her foreign trade in the decade from 1844 to 1854.*

As one speaks of the docks of London, another speaks of the Boston wharves. Today Boston has forty miles of waterfront; and of these forty, eight of them offer space for berthing on a depth of thirty

*Samuel E. Morison in his superb *Maritime History of Massachusetts* reports that "Boston Harbor never presented a more animated spectacle than during the clipper-ship era. One April day in 1854, wrote F. O. Dabney, no less than six large new clippers, undergoing the process of rigging, could be seen from his counting-room windows on Central Wharf. Across the harbor, the East Boston shore from Jeffries' Point to Chelsea Bridge was almost a continuous line of vessels in various stages of construction. Twenty ships of eleven hundred tons upward were built there that year."

to forty feet at mean low water. The world's largest ships may enter the central harbor which is served by almost seventy steamship lines. Being the chief American fishing port, as symbolized in the great Fish Pier, Boston joins her New England sister ports to make the northeast corner productive of more than one quarter of the nation's fish supply, which is (deducting scales and glue) some 500,000,000 pounds per year.

"And all I ask is a tall ship," said the present Poet Laureate of England. All that Boston asked was a tall ship with cloudy sails to give her the clipper-ship lead in the middle of the century that went out with Victoria.

"We can ask no more here [says Samuel E. Morison]. But in that unknown harbor toward which we all are scudding may our eyes behold some vision like that vouchsafed our fathers, when a California clipper ship made port after a voyage around the world.

"A summer day with a sea-turn in the wind. The Grand Banks fog, rolling in wave after wave, is dissolved by the perfumed breath of New England hayfields into a gentle haze, that turns the State House dome to old gold, films brick walls with a soft patina, and sifts blue shadows among the foliage of the Common elms. Out of the mist in Massachusetts Bay comes riding a clipper ship, with the effortless speed of an albatross. Her proud commander keeps skysails and studdingsails set past Boston light. After the long voyage she is in the pink of condition. Paintwork is spotless, decks holystoned cream-white, shrouds freshly tarred, ratlines square. Viewed through a powerful glass, her seizings, flemish-eyes, splices, and pointings are the perfection of the old-time art of rigging. The chafing-gear has just been removed, leaving spars and shrouds immaculate. The boys touched up her skysail poles with white paint, as she crossed the Bay. Boom-ending her studdingsails and hauling a few points on the wind to shoot the Narrows, between Georges and Gallups and Lovells Islands, she pays off again through President Road, and comes booming up the stream, a sight so beautiful that even the lounging soldiers at the Castle, persistent baiters of passing crews, are dumb with wonder and admiration."

It is wonderfully clear even to a landlubber like myself that those were the days. That full-rigged tracery against the seaward sky was surely something that no ocean commerce in this streamlined era has ever been able artistically to replace. But if the tall masts have long since fallen or been unstepped, the outstretched fingers of the great Boston wharves still carry the old romance in their length and look, in their queer old buildings, and in the lettering and words of their signs and advertisements. Say but the names of four: Rowe's Wharf, India, T, and Long! *There* is a quartet, chosen at random, to which tradition is safely anchored and which will mean the port of Boston as long as the old ship chandlery continues to punctuate Atlantic Avenue.

ANCESTORS' WORTH Not far up the Charles River from the Longfellow Bridge, the towers and spires of Harvard cut sharply into Cambridge sky. The oldest institution of higher learning in the United States, Harvard traces her proud history back to the year 1636. Her newer and Georgian front of red brick, ivy green, and variously colored domes sweeps gracefully with the curve of the river on the Cambridge side. On the Brighton bank opposite, round an open-front quadrangle, range the buildings of the Business School; and flanking them just a javelin throw upstream is the great level area of Soldiers Field, with its stadium, athletic buildings, and the adjacent Newell Boathouse. This is the Harvard with a view; the satisfying pattern comparable to the grey silhouette of Oxford from the London Road; of Gothic Princeton as the stranger beholds it from Princeton Junction; Amherst sighted from the east; or Dartmouth College rising hopefully on the hill a few miles up the way from Lebanon. This is the Harvard that the new aquatint would almost surely show us if anyone were making aquatints today.

But of course the heart of the University is the old College, and that lies to the north of the river, fenced in by the swirl and jangle of traffic which unfortunately makes Harvard Square one of the busiest arteries in the Greater Boston area. Here, however, withdrawn from all the noise and sick hurry of the day, the stranger may enter the Harvard Yard. Other American colleges have campuses, but Harvard has always had and always will have her Yard of grass and trees and youth and old familiar ghosts. There in the northwest corner, the oldest group of Harvard buildings links the fabulous past with the incredible present. These buildings are arranged, as the architectural fashion was, with "free form in Euclidean space." Massachusetts Hall, built at the public charge in 1720, has seen the nation and Harvard through ten wars. Designed as a residential hall, it quartered six hundred and forty soldiers during the Revolution. It was earlier the site of the first laboratory of experimental physics in America, and has served in turn as a dormitory, a theater workshop, and lately as an administration center, including the offices of the president.

The original Harvard Hall was burned in 1764, but a new one was built in 1766. A lecture hall today, it, too, has in other centuries given various services: the college kitchen was there; so was the buttery; so was the Chapel. Just beyond Harvard Hall is the little gem of Georgian architecture called Holden Chapel, built in 1744, gift of the widow of a rich London merchant. It has well been called "a solitary English daisy in a field of Yankee dandelions." Holden Chapel, queerly enough, served the College as a place of worship for only twenty years, after which it entered into a fantastic sequence of uses as barracks, senate chamber, courthouse, carpenter shop, engine house, dissecting theater, recitation building, museum, lecture hall, clubhouse, laboratory, and general auditorium. In that same corner of the Yard, though modernized with a sanitary bubbler, the old College Pump still functions in careful replica of the one long since destroyed. Its alleviating function today is a far cry from the frosty mornings of another age when the ambitious student —including Emerson himself—clattered down three flights of stairs in Hollis Hall to draw sufficient water for his modest ablutions. In our time this young man has a room, a study, a bath, and all the steam heat he can use.

As a university, democratic Harvard is somewhat of an empire. The little College for the training of ministers in the primitive colony has grown beyond the Yard into seven great upperclassmen houses, two of them named Eliot and Lowell, which stretch from just south of Massachusetts Avenue down to the river—each self-sufficient with commons, a handsome library of 10,000 volumes, for browsing as well as for study, and other facilities. Today the College, like the University, is completely national in its reach. Of the twelve thousand students enrolled as against the normal eighty-three or eighty-four hundred, about thirty per cent are from Massachusetts, thirty-eight and one half per cent from New England (including Massachusetts), and sixty-one and one half per cent from the rest of the country and the world at large, including representatives from every state in the Union.

I say the University has the structure of an empire. The great schools of Law and Business Administration are out there on the Charles; the Medical School is in Boston; the Arnold Arboretum is southward Jamaica-way; the Harvard Forest is up in Petersham; Harvard's oldest observatory is in Cambridge, but two others are, respectively, at Harvard, Massachusetts, and down in Bloemfontein, South Africa. The Botanical Gardens are located near Soledad, Cuba. Harvard has an interest in a station for the study of solar radiation at Climax, Colorado; and a partnership in the Yerkes laboratory of primate biology in Florida, for study of the giant apes. Harvard is weather-wise, too, with a Blue Hill Observatory and an interest in a new meteorological station on Mount Washington.

At the center of the College and the University is the University Library with a collection (as previously noted) of some five million books—the largest university library in the United States and probably in the world, exceeded in this country in size only by the Congressional in Washington. Its newest addition is the Lamont Library for the sole use of undergraduates. The Fogg Museum of Art, enriched in recent years by the great Winthrop collection, is today one of the foremost art centers in the land. Harvard's laboratories now sprawl far beyond the northern boundary of the Yard, completely surrounding that other museum forever linked with the name of Agassiz, wherein is housed for science and the aesthetic enjoyment of some 250,000 delighted visitors each year the priceless collection of the Blaschka glass flowers.

But material things alone have never made a college or a university. The greatness of either is measured ultimately in related terms of faculty, students, and that *tertium quid,* the alumni. Harvard's faculty is strong in the tradition of strength. Names of only yesterday, with magic in them, come to mind: Agassiz, Shaler, Norton, James, Story, Santayana, Royce, Briggs, Kittredge, Haskins, Channing, Perry, Copeland, Taussig, Lowes, and a hundred more. They have their able successors. The Olympian President Eliot, for forty years at the helm, and founder of the free elective system; President Lowell, who brought the University as distinct from the College to full flower, saw to completion a vast building program, and fathered the tutorial system and the House Plan —these two are succeeded by a third president of extraordinary vision and versatility in James Bryant Conant.

Harvard's sons have played a distinguished part in the building of a democracy since long before that democracy was clearly envisioned. She has given the nation four presidents—two Adamses and two Roosevelts—and leaders on every frontier of human endeavor. Names of the past like Mather, Emerson, Thoreau, Parkman, Holmes are hers. Hers, too, are Walter Lippmann, Robert Sherwood, Robert Frost, Edwin Arlington Robinson, T. S. Eliot, Van Wyck Brooks, Bernard De Voto, and Samuel Eliot Morison. These names are representative in the field of letters; but they are duplicated in the field of science, medicine, law, business, government, art, music, and education.

In 1936 the world of scholars gathered in Cambridge to celebrate Harvard's Tercentenary. It was a dynamic spectacle of color and excitement, but it was also an occasion of deep humility and consideration of the homespun past and dangerous days gone by. The world at the moment was preparing for war. Like other colleges and universities across the face of the land, today Harvard is preparing for peace. She is one of the great vitalizing forces in our community; her name is known and respected throughout the civilized world. She is with us till the stock of the Puritans die.

THE WIND THAT BLOWS Few people, I gather, have much regard for the signs of the zodiac through which the sun passes in the twelve divisions of the year. Not many of us can tell, to my sorrow, the sign under which we were born. This seems to me a pity, for the night sky is full of comforting symbols, and one of them is our own. Gemini or Scorpio, the Fishes or the Lion, one of them is something personal and familiar to look for and remember—the skymark you carry with you when you are far from home. The star fields never change: their paths are the same: as great in their loneliness as we are small in ours.

Every twenty-first of March begins the month-long reign of Aries the Ram, but we don't have to have been born under his aegis to know that his arrival means the arrival of spring. We think that the countryman is the fellow who really knows when winter is over—that the news filters into the city only after the deal has been put through. Some cities find it out by waking one morning awash in the overflow of the great river at its doorstep.

Others seem to discover it largely by the character of the millinery in the shopwindows and on passing heads. But a few are fortunate in having the country always in sight. Boston is one of this favored group; and the smell of spring, when it comes, is just as evident on Boylston Street as it is in the Merrimac Valley.

Boston even prepares for spring. Along in February or March the ice goes out—not with a roar the way it does down East, but quietly as a tear slips from the eye. Gutters reappear, and men with a morbid concern for what goes on under manholes remove the lids, survey the unattractive, and reach for the unreachable. The brightness of flowers behind glass or banked in certain conservatory windows, or sold in the open alleyway or under the shadow of a church, is echoed now by the yellow crocus and his purple brother, glad of the chance to be up and out. The east wind full of the sea, the west wind full of the Berkshires, the north wind full of Monadnock, and the south wind full of disappointment share in this preparation. Hats blow off, the traffic blossoms with new cars, lunch hours are stretched, the lavender and catnip vendors move into the sun, dogs strain at the leash, and the snow trains slink out of the North Station. Men with red balloons and multicolored pinwheels appear from nowhere, as round the corner comes the sound of an ancient tune ground from an ancient box with all the happy discord we had half forgotten.

Brass knobs and brass numerals on doors, and on the iron railings and grillwork so much a part of the Boston façade, shine from a recent polishing. Brighter still, the myriad brass plates, suggesting that every tenth Bostonian is a doctor, flash again in the sunlight. One could arrive at a very good likeness of Boston simply by rubbing its face of brass. Over in the Public Garden the wind is warm; the great willows by the pond show yellow in the branch; the magic pond itself is drained, as workmen rake and tidy up its disillusioning bottom. The sound of a hammer!—and that would be the carpenters laying new planks on the swan-boat landing. Somewhere the swan boats themselves are in dry dock, preening their plumage in anticipation. For a day or two now on almost any westbound street you may pass a truckload or two of city twigs and prunings fresher than some of our imports.

From the basin of the Charles River the last vestige of floating ice has melted, and the cloud of sea gulls which used to gather in hundreds, if not in thousands, at the edge of open water suddenly takes to the sky. Boston is a city of gulls. The wind blows them over the Common, across the Hill, above the few high buildings, and out to the harbor. Or if the day is still, the first March thermal bears them upward in their long and endless circling into the blue. The old gull with a wary glance who feeds at my fifth-story window sits on the chimney top, ruffles his feathers, and eyes the inferior pigeons which, in a moment, he will majestically disperse.

These yellow crocuses and the earlier snowdrops which punctuate the tiny front plots of ground dear to the Boston householder are, of course, not wholly reliable as prophets. They are apt to unfold to imperfect conclusions. One of them to my recent knowledge sang his solo in the Back Bay under the reign of Pisces. He paid for it dearly with his saffron life. The perennial mass appearance of the crocus may again be exhibitionism; but it reminds us of the flowering of the magnolias certain to follow. When they arrive in street and avenue, the city has nothing further to fear in the elements that the east wind or a warm rain cannot cure.

I say you can smell the spring in Boston because you are as close to the country as you are to the sea. You wake to the smell of it one morning shortly after daylight when it blows in through the open window and will not let you close your eyes. But before that—before even the boldest crocus—comes the black crow with the sound of spring. You have not heard him much through the winter, when he is quiet about his affairs; but suddenly his city voice is rising above the rooftops, and you stop in your tracks to listen. I have seen him already on the ledge there, where my gull descended, examining an old green glass insulator on my neighbor's fallen aerial. His voice is not less national than the radio, and he is somewhat more of the philosopher. I can hear him now in fancy,

> Saying once more:
> March is outside the door
> Flaming some old desire
> As man turns uneasily from his fire.

Well, the Bostonian turning uneasily from his fire has all the New England countryside, the two Capes, the Maine coast and the Maine woods, and three divisions of the Appalachian Mountains almost at his door. But he has something even better because it is more immediate: a greater area of grassroots and shrubs and trees and flowers in his front yard than in the heart of almost any other American city of our latitude. His Common, the

Public Garden, and the Esplanade on their made land provide his link with the outer landscape. That's where his spring unfolds. That's where the green begins. That's where his city is youngest, because it is always fresh in contrast. Crossing the Common tomorrow, and every tomorrow, he may well think as Thoreau thought at Walden:

> Men say they know many things;
> But lo! they have taken wings,—
>
> The arts and sciences,
> And a thousand appliances:
>
> The wind that blows
> Is all that anybody knows.

DR. HOLMES'S GARDEN A stranger, walking west across the Boston Common to Charles Street and so on through the Public Garden, must surely imagine that the Common and the Garden are a unity which somehow became divided by a street. But for all that suggested continuity of grass and trees, such is not the case. Indeed, the Common has something more than one hundred and fifty years of exciting and even stormy history behind it before the area which we know as the Public Garden was anything better than the so-called Round Marsh—or the marsh at the bottom of the Common. What caused the transformation? There is a bit of luck in the story of its evolution.

By the close of the eighteenth century, rope-making was a recognized Boston industry; and when in 1794 a number of ropewalks on Pearl and Atkinson (now Congress) streets were burned, a meeting of the people was held to determine "whether the town would appropriate the marsh at the bottom of the Common or any other of the town lands for ropewalks for the accommodation of the sufferers by the late fire." The upshot was that a committee made a survey of the marsh "sufficient for the purpose of erecting as many ropewalks" as had been consumed. The town, with one eye to controlling the filled-in marshland and the other shining with the light of simple charity, granted to the ropemakers the right to build not more than six ropewalks one story in height. The grantees, in turn, were to build within two years a sea wall along the whole length of the new land, which must have cut north and south across what is now the middle of the Garden. Though each of the ropemakers was supposed to be offered a deed to his plot, only one deed was actually given. The new ropewalks, had you been there to see them, stretched about three quarters of the distance along the west side of Charles Street from what is now Park Square toward Beacon Street.

Unfortunately, the new ropewalks burned down in 1806. The optimistic ropemakers, however, now decided not to raise another phoenix from the ashes, but to cut up the land into lots to be sold for building and dwelling purposes. Charles Street itself, having been opened in 1804, had obviously enhanced the new land value. But the people of the town of Boston, mindful of the unobstructed view of waste and water out where Commonwealth Avenue stretches today, strongly objected. In the end, so queer is fate and the gambling instinct of man, the town itself was roped and thrown by the enterprising group and obliged to pay some $54,000 to regain the rights to the acreage it had given away just thirty years before! The agitation to build continued from other quarters until 1859, when by act of the legislature and vote of the city the twenty-four acres as we know and use them today were given forever to the people as a Public Garden.

Back in those ropewalk days, presumably just inside the present center gate of the Public Garden on the Charles Street side, a slight elevation was known as Fox Hill. In the proper tide it became an island, and it was just large enough to have been fortified by the British in 1775. Its greater and ultimate usefulness, however, was demonstrated by the town, which leveled it to cover the surrounding bog. A load or two of oyster shells were thrown in for good measure. In 1839 a few citizens of vision laid out a small botanic garden in the favored area near the corner of Beacon and Charles streets. The venture flourished until the conservatory—like the ropewalks before it—was destroyed by fire. Twenty years later, when traveling circuses camped on the Garden grounds, the twenty-four acres were duly landscaped. In 1861 some other thoughtful souls added the pond of four acres—a pond which contented itself with the water overflowing from the most famous frog pond in the world, lying but a few hundred yards to the east. It was and still is spanned by a stone-and-iron bridge beneath which glides the quintet or so of swan boats loaded to the gunwales with children between the ages of two and eighty-two. Now, on the first of May, punctual as the returning swallow, thousands of tulips in the Garden suddenly burst into bloom, the many saucer magnolias open their cups, the sketching classes of the lately hibernated art schools emerge with pencil and brush, perambulators appear, the various park benches begin to fill, and an old affection between the city and the country is renewed.

These northern city gardens, like ours and that

other one far to the east in Halifax, provide us with a little firsthand knowledge about the horticulture of other and more southern countries. With the summer come the potted palms and such. But all through the year the stroller with any curiosity at all can learn merely by reading some well-weathered labels the names of many trees with which the New England farmer is unfamiliar. Of course we are not surprised to find *Ulmus Americana*—the American elm—along the mall on the Beacon Street side. And weeping willows, rock maples, the horse chestnut, and even *Fagus Grandifolia*—our American beech—are likewise familiar enough. But how about that noble tree, the giant Van Geert poplar, at the pond's edge? Or the lovely tea crab with its rosy young bud? Or this queerly twisted, arthritic Camperdown elm of the liquid name? Unless your mind steers like a compass on the shortest haul from Arlington to Charles Street, you are more than likely to bump into a purple beech, a Holland elm, a Scotch elm, a European ash, or that solid old customer, the mossy-cup oak. And of course the ginkgo trees of the fanlike leaf which Marco Polo saw in China centuries ago.

In 1862 an iron fence replaced the wooden one round what Dr. Holmes in the *Autocrat* refers to as "my Garden." But one class of visitor there has never known the limitations of a fence. The old records are full of the names of local birds and birds of passage observed in the branches of some of the trees of which I speak. Warblers, thrushes, vireos, cedar waxwings, redpolls, vesper and other sparrows, kingbirds, siskins, flycatchers, crossbills—the list is endless even down to recent years. One night in April I entered my door just seven blocks from the Public Garden to find in my library an olive-backed thrush. He flew across the room and perched on the frame of my Benson wash drawing of "Crows in Winter." It was a male bird—the males are apt to precede the females journeying north. Whether or not he had missed the Garden in the storm I cannot say; but in releasing him the following afternoon I was reminded again of the wisdom of the founding fathers who gave this city so beautiful a landing field so long ago.

NORTH END Ask the stranger within our gates for Boston's second name and he will probably reply Back Bay. It is queer how this alliterative tag for the most recent man-made land of the city has grown in popularity until it means far more than the tributaries of Commonwealth Avenue. Even the Hub as a place-word has been eclipsed. But there is another name for the oldest section of Boston which still holds its own. This, of course, is the North End. No one will deny the magic in those two monosyllables, simple, abrupt, and unpoetic as they are when either written or spoken. What a stockpile of history they now control! It needs no metallic equestrian Paul Revere in the mall that bears his name to give us the most familiar symbol of all. Nor the image of his diminutive house at Number 19 North Square, which has survived in this world of demolition since 1677 and which Paul purchased in 1770. Nor the sight of the spire of Christ Church; nor the other-worldliness of the weathered headstones long since planted in the Old North Burying Ground on Copps Hill on Hull Street.

The North End is more than the sum of these tangibles. It is, rather, a legend familiar to every American schoolboy, whether he has set eyes on a single one of its twisted streets or not.

And what is the North End? For the Philadelphian heading from Penn's Wood for a lazy summer on the fog-bound coast of Maine, it is the obstacle which separates for him, by the measure of a taxi fare, the South Station from the North. It is the foreign region which invites the householder in search of fresh fish and lobsters; of provolone cheese and French bread; of all the vegetables, native and obscure, from asparagus to zucchini; of local fruits and imported. It is the tourist mecca, attracting 80.000 visitors in a single summer to the Old North Church alone. It is home to the great Italian population of Boston where, in the compass of a few blocks, there is an average of 1,300 persons to the acre—the most congested area in the world except for that other one in Cairo, Egypt. It is the haven of him who perambulates in search of Joe Malatesta's or other fine dispensers of spaghetti. It is the provider of the shortest arterial route from the waterfront to Scollay Square for those gentlemen of infinite leisure who become alerted only in the hours after dark. It is a labyrinth of the most parabolic and curious streets in a naturally curvaceous city. It is the secondhand depot of the cheap, the twilight of the tawdry, the flowering of the wholesale, the noisy market place of a million, and the safe harbor of much that is very lovable and authentically of the past.

Geographically the North End is that part of Boston which lies on the polestar side of State, Court, and Cambridge streets. Once "the wealthiest, most populous, and in every way the most important part of the town," it has long since lost the first and last of these three impressive titles. From an area of more than three hundred acres, rich in association with the Crown, republican sentiment has erased many of the old and aromatic designations. King Street is now State Street, Queen has lost to Court, though Hanover for some reason still prevails. Delightful, fragrant names like Gay Alley, Paddy's Alley, Moon Street Court, Beer Lane, Black Horse Lane, Back Alley (so narrow a drunken man could not fall either to the right or left), the Old Way, and such, have been exchanged by the unimaginative and the practical for the presently familiar, or have simply disappeared without replacement. Before the numbering of North End streets, in the day of Benjamin Franklin's father, whose shop sign was a blue ball suspended by an iron rod over the southeast corner of Hanover and Union streets, it was customary to designate one's place of business by an emblem. To the sorrow of the antiquary, we have lost through progress such emblematic names as the Heart and Crown, Three Nuns and a Comb, Brazen Head, Tun and Bacchus, Three Sugar Loaves and Canister. To our credit we still have on Marshall Street at the back of Joe Malatesta's restaurant and opposite the oldest brick building in Boston the Boston Stone bearing the graven date of 1737. It took its name from the London Stone; was brought from England in 1700, originally used as a paint mill by the painter who then occupied the Marshall Street premises, and appears later to have been considered as a starting point for surveyors and as a direction marker for the local shops. Probably not a dozen of the readers of these words have ever seen it.

In the long stretch since the landing of the first settlers somewhere near the foot of Copps Hill, through the years when by virtue of creek and canal part of the North End was practically an island, and Mill Pond an unlikely candidate for the site of the coming Boston and Maine Railroad Station—through more than three centuries, in fact, this heart and center of old Boston has known many tenants. In the last century alone it has experienced (not without conflict) a change of population from Yankee to Irish to Irish-Jewish-Italian and ultimately to Italian.

In the middle of the North End today, and considerably a landmark in its own right, is the North Bennet Street Industrial School, founded in 1881 by Pauline Agassiz Shaw. Its devoted director is Mr. George C. Greener. In addition to the school's chief function of giving industrial education, it is a living influence for improvement in a highly congested tenement district. It provides children with training in good habits, adolescents with guidance, adults with social and economic adjustments, and people of all ages with tangible community service. It is a kind of beacon on a hill of its own.

The annual flow of visitors to the North End is largely a tide of out-of-town and out-of-state people. Yet the native Bostonian, in his quiet, untalkative way, is fast coming to realize the commercial as well as the spiritual value of the old landmarks, so many of which are spread upon this febrile acre. He knows that they should and must be preserved; that these old buildings—some of them of English brick and foot-square timbers cut from the vicinity of

Copps Hill itself—need protection from the vandal as well as from the demolisher; that the city government, already actively interested, should have a hand in their saving and restoration. The North End! A queer composite, this slope of land, with its dockside teeth biting into the harbor and the very hide of it bristling with dark red clay and chimney pots. This is the real Boston; this is America, old and new. And it is rather comforting to remember that here in the house at the end of Sheafe Street once lived the author of some verses beginning "My country, 'tis of thee."

QUINEBOQUIN Cities and rivers enjoy one of the oldest relationships in the history of civilized man. Nineveh on the bank of the Tigris is a ruin, but Cairo on the Nile is not. If one of the two has to go, it is always what man has wrought and never what is a part of nature. Rivers change, to be sure, and you have what the geologists call drowned rivers—like the Hudson—where the original mouth is now far out under the sea. But changes like that occur in geologic time; whereas changes in cities occur in man's time, which in the geologic sense is and must always be brief.

The reasons why so many cities great and small are founded on the banks of rivers are obvious. Water, transportation, power, agriculture, security, and defense are some of them. Scenically and practically they are good for each other. Rivers and cities have a way of mingling their identity. To think of one is often tantamount to thinking of the other. London on the Thames is an example; Paris on the Seine, Rome on the Tiber, Quebec on the St. Lawrence—the list is endless. Some cities stand at the confluence of two rivers, like Pittsburgh; others are near the confluence, like Omaha, St. Louis, and Portland, Oregon. New York is one of the few great cities of the world situated exactly at the mouth of a large and navigable river; and it stands on solid ground. Pará, for instance, in the complicated mouth of the mighty Amazon, and Astoria at the mouth of the beautiful Columbia are relatively small cities. In the case of the big rivers, the shifting deltas at the outlet, like the muddy delta of the Mississippi below New Orleans, argue insecurity, and man seems to have said to himself: Let's go upstream a little way. And he usually did.

Boston, a relatively large city, stands at the mouth of a small river. A river small in size and volume, that is, but not small in the history of Massachusetts. The Algonquin Indian name for it was *Quineboquin,* which means twisting or winding. But to the eyes of Champlain some three hundred and forty-odd years ago it appeared to be a very broad river; and so it would seem to anyone today sailing into the harbor, which resembles the business end of a stream quite as much as it does the sheltered inlet of the sea. But Champlain apparently never got as far as the first bend. Even Captain John Smith, who followed him, also took the landward stretch of it largely for granted. Of all the weary rivers of which Swinburne was to sing, the little Charles in its oxbows and coils, in its losing argument with a glaciated soil, in its vague and indecisive effort to find the open sea, is a queer one indeed to have fooled the early voyagers with that thimbleful of water and grave, unhurried pace. To the visiting motorist in the area west and southwest of Boston, the Charles turns up in a surprising number of towns. Take a look at it up near Hopkinton—the fountainhead. It bears little promise, doesn't it, of reaching *both* Bellingham and Millis? Cross it in South Natick at the site of the old Indian bridge and you will give it an equally faint chance of arriving at Waltham by way of Dedham. To the Bostonian his river is at its best in the stretch from the Basin dam up to Watertown. He will not remind you, a stranger, that one hundred years ago the water lapped at the western edge of Arlington Street, and that within his memory there was once no Basin with bright sails and creaking shells upon it. In the memory of his grandfather, one could swim off Charles Street just downstream from Mt. Vernon. What is now the Basin was marshland, and a rather sorry ending for Charles the Sinuous—tidal clear to Watertown. Shad and alewives swam upstream in season—through what was once called the Upper Marsh—and the fish weir at Watertown is said to have taken as many as a hundred thousand alewives in two tides. It is well to remember, too, that the first and so-called "Great Bridge" across the Charles between Cambridge and Brighton (near the site of the present Anderson Bridge) was completed in 1662, and that for thirty-two years previous the Boston-Charlestown ferry was the *only* Boston crossing. It was this ferry, incidentally, which for a century was operated by Harvard, the rent therefrom being, in the ungilded phrase, "a considerable part of the support of the College." With muffled oars, Paul Revere was rowed across the Charles to the Charlestown shore one exciting April night; and twenty-two years later,* at Edmund Hartt's Shipyard, just where the river emptied into the sea,

*October, 1797.

and not far from Paul Revere's own house, the *Constitution* was launched after a month of thwarted attempts to get her down the ways.

Today the Charles at Boston flows by, unmindful of these things. Indeed, it does not appear actually to flow at all. It is more like a great mirror held to the city's most favoring profile. Goldfish, carp, the feral ducks, and countless restive gulls—they know it best. At the lower end there is now and then some traffic in cargoes up the two brief canals in the direction of industrial Cambridge—or down and out through the lock to the harbor. In summer the several small excursion boats again make the grand tour. Some of them, like the *Elk,* will be vanishing with their tourist and native passengers in the direction of Watertown—a long cool voyage on a wretched hot night. From the height of any railing it is good water to look at. The sun shines on it, the rain falls into it, the gales raise some angry-looking whitecaps, the west wind rinses the sky, and smoke-blue mist of evening softens the old city at its elbow. At night the lights, like stars and floating oranges, color the backdrop of the stage, and a train drifting silently out of the subway across the pepperpot bridge has the look of a tired old comet half ablaze—and half asleep.

2. A learned teacher of law tells me that some years ago he was driving out in the direction of Dover with his eight-year-old son. In the course of their drive they kept passing and crossing a capricious stream of water. They passed and crossed it in towns, in the country, by old bridges and new. The youngster at intervals would ask: "What river

is this?" The answer was always: "The Charles." Finally they crossed it one time too many—flowing, as usual, in a fresh direction. The same answer excited the small boy to say: "Well, why doesn't it make up its mind?"

A number of my boyhood years were spent in the southwest of the state of Oregon on the wild and talkative Rogue River—one of the great home waters of the Chinook salmon and the steelhead. It is a fast river; falling, I think, some twelve feet to the mile. What the Charles does at Newton Upper Falls and Newton Lower Falls, the Rogue does at every third bend—and it, too, is uncial and wayward. But because the Charles is unambitious—not very useful, not navigable above the Longfellow Bridge for anything much larger than a police boat—it brings to Boston from opposite directions an older and better reminder of the sea and an ancient symbol of the country. It opens to the city both the front door and the back—not to boats and barges steaming in, or to logs coming down, as you may find them on some other American rivers—but to life on the wing, or at least to that part of it on the wing for which a placid river is at once an attraction, a source of food, an anchorage, and a safeguard. It is this very sleepiness of the lower Charles that fascinates me now, as the swiftness of my boyhood Rogue enchanted my restless spirit then.

Against the current we have the gulls. I watch them through the winter, gathered in midstream on the ice. On the coldest day they sit there facing exactly into the wind, but streamlined against it, for gulls are perhaps the most sensitive of living weather vanes. This is their welcome mooring after the morning forage on distant dumps. Along in spring they flock to the edges of the open water, some of them in the water; and they move downstream together with the passing days until the disappearance of the last of the frozen surface. Against the solid and dirty ice they look in the distance like a drift of clear fresh snow. Spring comes in as the gulls go out. Summer takes them to the harbor and thence to sea; but there is really never a day when some of them aren't winging back over or circling the wide area of the Basin.

As the ice goes out, and for some days thereafter, driftwood and strange objects from the Watertown littoral glide slowly down. Small boys begin to fish for shiners. With the current come other birds to replace the gulls. Everyone who walks the Esplanade in spring and summer is aware of the numerous pairs of mallard ducks which cruise inshore from the inlet up to the mouth of Muddy River and beyond, looking for peanuts and other gratuitous edibles. They nest in secret almost within reach of the hands of those who feed them, and the youngsters surviving the predatory rats, and turtles, likely, and all the associated threats of civilization bicycle after their dark-feathered mother in the shallows, rising and falling with the mimic waves from passing boats or those created by the inshore winds. They used to nest along the tiny cattail-blackbird ditches of what is or was left of the Lower Marsh—not far downstream from the Western Avenue Bridge. It was pathetic in the 1930s to watch there a convoy of seven youngsters dwindle day by day to two or three—usually motherless at that stage, thanks to natural or unnatural enemies. One can find all the heartbreak of the northern woods in the proud wild look of a single green-ringed mallard drake.

In the dead of winter I have watched the feral sheldrakes—like white chips bobbing in the dark of the open water below the bridge piers—swimming often in the company of mallards at the edge of the ice cake on which the gulls were assembled. One rainy day when the wind was blowing hard and the rain falling thinly but fast, I observed a queer performance. I can't think of another riverside city where it could have been so easily observed. Three gulls were trying to maneuver on the slippery windswept frozen surface:

> If a gull swung round, he would take a brace
> With his wings out full to keep his place—
> Like a child who walks with a want of grace.
>
> If he took one step he began to slide
> Or he sat right down in his wounded pride,
> With his wings still poised and dignified.
>
> It was clear at least that a graceful bird,
> With more in his flight than the poet's word,
> Has a way of his own to appear absurd:
>
> His following ships and the shoreless sound
> In freedom hardly the sky can bound
> Is a lonely humor when gone aground.

But gulls and mallards are not all. Within a year or two I have come upon a bittern standing just upstream from the railroad bridge, while a Boston & Albany flyer was thundering by. Not long ago two hooded mergansers were there for a day. I have seen a kingfisher, in Henry Williamson's great phrase, draw a blue line between two bridges, and my pair of famished hawks each spring return to hunt the mouse-filled vestiges of marshland where a golf range has been landscaped against the summer. I followed one of them one day in my car, driving him from tree to tree in the stretch above the Anderson Bridge. He had a mouse in his talons. Just at the great bend which looks up toward Watertown I lately came upon some forty redpolls tumbling like champagne bubbles in a little leafless birch. And on another spring day out of nature's sample case, I found a solitary black-crowned night heron presiding hopefully over a frogless area about as far from Braves Field as a confident man could possibly knock a homer.

But life on the lower Charles is not all riparian either. Even a rainy morning in the spring months will reward the passer-by with sight of three or four eights and a motherly coaching launch out for improvement of form and strength in the blade. The blend of sweaters of the young men who do the rowing is almost as decorative as my night heron's crest, and the batrachian sound from the coxswain's megaphone floats shoreward like the vanished croakings from the now departed marsh. Well-filled white sails from the Boston and Technology shores replace the sheldrakes; and one adventurous individual who slips upstream and down all summer in his private kayak, to the rhythmic twinkle of his double paddle, is the low-compression counterpart of my lonely friend the kingfisher. Boston does its quiet breathing on the Charles.

THREE IF BY AIR It is deceptive to know a city only by day. You can get the look and loom of it best, of course, in the daylight hours, but the temper of its quality, whatever it has of somberness or gaiety, its rhythm, and the measure of its greatness on the metropolitan, if not the cosmopolitan, scale are never quite so clear under the sun as under the private light which man has devised and given it. A fine city, like a good play, can be spoiled by bad lighting.

On the other hand, for some people a city is most reflecting and reflective at its awakening. Marketmen and milkmen, I always think, have the best hours of it to themselves. The weary nightworker plodding home with heavy eyes is perhaps as unaware of this morning magic as a robin would be on the owl-shift in pursuit of glowworms. But those who are by habit up with the dawn and 'ware and wakin', as Sir Henry Newbolt says in the best stanza of *Drake's Drum,* are members of a lucky race. They may not be conscious of this luck, and perhaps a Gallup poll would prove me wrong. Custom for them may have staled the replenishing miracle that poets write about. But certainly for the rest of us there is something in merely driving out of a city in the very early morning that is like beginning a new life or heading to sea for the voyage we always intend to take. Familiar streets and houses seem fresh and new and vaguely unfamiliar. A city is handsome and impossibly clean at that daybreak quiet hour.

Now fishing is an obsession that gets a few of us up before the sun at least once or twice of a spring or summer morning. Several years ago (late at night) I wrote a sonnet about such an occurrence. An odd pragmatist, I called it for what it was:

Early For Once

At five o'clock in the morning clear
On the one sweet day for the blameless end,
When rising early I seem to hear
In the birds of the city the country blend;
And gathering tackle and rods and pack,
I slip downstairs and unlatch the door,
The dawn has opened the wider crack
And the stars have lessened to three or four.
But nobody walks in the timeless street:
From far away comes a muffled sound.
The day that my heart has cried "Retreat!"
I seem to be quitting the lonelier ground
For a boyhood brook in a depth of wood
That is close at hand, if I understood.

Of course I used the words "if I understood" to mean not literally that there might still be a fishing brook in Boston, but that there is in any old city like ours no end of undiscovered, secret places of enchantment equivalent to some of the simple but lost things of our own lost youth. These places seem nearest at hand—seem almost round the corner—when the pavements are grey and empty. The welcome loneliness to which one surrenders in the abnormal pre-breakfast desolation of certain Boston streets can suddenly suggest that the country by contrast is an undesirable, noisy kingdom far away. Even the grimy synoptic look is gone for a moment, since the light of early morning has its own way of washing the face of brick and stone.

But I have been talking of what most of us—certainly myself included—but rarely see. Evening and night are another matter. Henry Beston, in a beautiful Cape Cod book called *The Outermost House,* has suggested that man is getting out of touch with night. He means, of course, the country night of solitude and stars. That, alas! we can agree is true; but the aspect of cities after nightfall is something with which man is becoming increasingly familiar. Indeed, the average city dweller, like the moth with respect to the flame, finds some aesthetic satisfaction in the kaleidoscopic pattern of his neon background. By the yardstick of other large American cities, Boston is neither underlighted nor overlighted. She is comfortably bright without being gaudy.

With the coming of dusk the mystery of a city deepens. Looking across the Charles toward that apiary Hill fading from brick dust to blue to the colorless substance of shadow, with the few tall buildings standing out in final relief, I have many times watched the lights come on. Below the twinkle of windows, the Esplanade defines itself in large round yellow periods. A few bright signs aloft hotels and the well-spaced radiance of floodlights from

Bunker Hill to Federal Street unfold with casual independence. Palest and earliest, the ghost-green cap of the New England Mutual assumes the glow of willemite against the sky. South of it John Hancock's improved Empire State Building looms just big enough to create a backdrop and make Mr. Cram's odd structure resemble a Pushman Buddha, squat and mysterious, lacking only the expected tear vase and fallen petal. From the middle of the Longfellow and Harvard bridges to the Boston shoreline appear four strings of lamps. A few minutes later economical Cambridge responds by closing the circuit for the corresponding halves. Out toward the Newtons and down to the sea the whole of Boston is alerted.

On a dark, wet night those bridge festoons furnish all the radiance one sees in driving to town on the river side from Brighton, and they often resemble two great hawsers mooring Mr. Blaxton's old peninsula to the mainland. On cloudy nights Boston shows herself as a red glow in the sky. In the blackout nights of the war she reverted impressively to the city of her ancestors—not only full of shadows and Colonial ghosts, but silent also. She had a peculiar beauty then which many remarked and remember. It was easy for once to understand how "one if by land and two if by sea" could be visible in Charlestown from beyond Copps Hill.

In her quiet way, night by night, Boston grows noticeably luminous, like the modernized dial of an inherited watch. Huntington Avenue now flows brightly into brightness. Where yesterday we had but one ball park with its own solar system, today we have two. On her Back Bay skyline the new John Hancock affair, its observation tower tentative with red, will sponsor a nova. But there is nowhere as yet sufficient aerial dazzle to mar the fascinating pin-point patterns of the waterfront, the candy-stick columns reflected in the upper Basin of the Charles, the romantic reds and purples of the railroad yards by the North Station, and the dwarf lights along the many tracks in many directions that link us to the world. After sundown those links are with distant places only—the country in between has been swallowed up.

> Such flowers strewn by night
> At the switch's groin
> Are Idaho, Seabright,
> Quebec, Des Moines.

And of course the great highways of the sky have their lights as well. "Three if by air" would have sounded odd to Paul Revere. But his triple (not twin) lanterns hang there in red tonight and every night, just a little higher than they could have hung in the steeple he called North Church.

NOW AND THEN A city is no better or worse in character than its streets. The finest buildings in the world lose their beauty of profile, their impressiveness, and much of their outward attraction if they are poorly located in an uninviting street. Now the length and width of a city's high- and byways have little or nothing to do with our judgment of them as to charm and personality. The meanest street in the world is not necessarily without interest—far from it; and the widest and longest is not necessarily more exciting or romantic than a block or two of some queer and crooked alley which has weathered a few centuries and is, by the Whistler standard of all great art, quite low in tone. The great avenues of the world have a certain lift to them, the like of which a hundred thousand lesser streets cannot claim to offer. To walk up Fifth Avenue in New York City on an early fall afternoon when the flags are all flying in the breeze and the windows gleam, and the air smells of the sea, and the great shopwindows are at their best, is an experience to which even the dullest among us must respond with a sudden quickening of the pulse. The sight from Forty-seventh Street north on such a day is something to be remembered for a lifetime. So it is with the view up the Champs-Élysées toward the Arc de Triomphe in Paris, and with the other great boulevards of the Londons and Viennas of every continent. I think sometimes that we remember a city which we have visited largely by some single street that took our fancy. For my part, I shall always remember San Francisco, first seen in my twelfth year, by the cable cars of the perpendicular street called Sacramento; New Orleans by Royal Street and its offshoots in the Vieux Carré; Portland, Oregon, by the roads and drives on Portland Heights on up to Council Crest; and Quebec by Dufferin Terrace, which is an infinite platform kind of street; and by Sous-le-cap, a fascinating crevice through which, under crisscrossed airing laundry of bright colors, you try to squeeze your way.

I say all this by way of preface to some personal observations on the streets of Boston, because streets, much more than cities as a whole, invite comparison with one another. Years ago I put in the margin of some notes on Paris that all good streets are parabolic. I believe this to be true. There is something about the curving of any path, urban or otherwise, that is more inviting to the pedestrian or traveler than the way that is straight. "Two roads diverged in a yellow wood," said Robert Frost, and in that very divergence is the suggestion of the happy sadness of not being able to see round either bend. So with much of this city. The streets of older Boston are like the contours of a map of a mildly hilly region. They turn and twist on each other without provocation. It is undoubtedly harder for a stranger to learn his way about in downtown Boston than in any other American metropolis. The streets conform to the old cow paths, or so tradition says; the buildings conform to the streets by concavities and convexities vastly pleasing to the eye; it is only the citizen afoot and the stranger in the car with the Kansas license plates who have difficulty in conforming to the crooked result. When one reflects on the regrettable disappearance of such names as Gay Alley, Pudding Lane, and Black Horse Lane it is because one likes the sound of their syllables. But a change of name has rarely changed the direction of a Boston street, and the city is still full of queer byways, passages, and alleys, whatever they may be called. Have you ever considered Quaker Lane at the head of State Street? I cannot remember another anywhere which so quickly satisfies Robert Browning's "It's a long lane that knows no turnings." Quaker Lane, if you examine it, runs in five directions at once. And then, where will you find an avenue as short as Franklin? I do not mean Franklin Street, but Franklin Avenue, which is four paces wide and twenty paces—including nine steps down—in length, covering the incredible distance between Cornhill and Brattle Street. Cornhill itself is my idea of the aesthetically curved street—something like a comma set against the Bodoni, Bembo, and Cloister of Cambridge, Court, and State. Franklin Street tapers off much like a whiplash. Remembering the charm of old houses wherein two or three rooms on the ground floor may be at a slightly different level, I sometimes like to walk to the end of Bosworth Street just for the pleasure of descending those few steps into Province. And one of the delights of wandering through the downtown and North End labyrinth is to seek out the hidden passageways leading from one street to another, especially in the market district; and to discover, even for the twentieth time, the sudden delight in emerging from dismal High Street into the bright green of Fort Hill Square.

A man could spend years tracing the origin of some of the oldest streets of Boston. Back of the day when the North End was defined on its eastern axis by Salem, Hanover, and North streets, the configuration of tidal water, Mill Creek, and its extension as part of the Middlesex Canal is no clear blueprint to the arteries and veins of today. Blackstone Street, for example, named for the founder of Boston, follows roughly the bed of that canal. Streets of the old city have been lost and found and lost again. Some appear to be but fragments of themselves. We have West Cedar Street on the Hill, but there is no Cedar or East Cedar to go with it. Newbury Street, once the name for busy Washington, is now lodged in the Back Bay. Somewhere out toward the west, the Fenway cuts it off completely, but you will find it again on the other side, full of life and cheerfully emerging into Brookline Avenue.

Everyone knows that the great street of Edinburgh is Princes Street, which Glasgow—possibly in envy—calls half a street. Boston is full of half streets too: Beacon in its finest segment, a slice of Boylston and Arlington; and the faded flower of Tremont, defining the south side of the Common with a graceful arc. In a city of bay windows, this perceptible swelling of Tremont is a bay window in itself. In competition with the world's great avenues, Boston has Commonwealth: long in its ultimate progress through metropolitan districts, but brief and beautiful largely in its approach to the Public Garden. Width and trees and the central mall make it memorable; but it has architecture, too, of "ambitious variety" in period, style, and stone, and the good fortune to be intersected alphabetically by streets of alternate three-syllable and two-syllable names—Berkeley and Clarendon—as noble in sound as the nobility from which they stem.

And as to that, which American city can point to so many lovely, queer, and compelling names for its streets as Boston? Beacon, Batterymarch, Pinckney, Tremont, Hereford, Salutation, Merchant's Row, Cornhill, Brick Alley, and Sun Court Street, for example. And who else lives where Winter runs forever into Summer, and Water is always on the verge of turning into Milk?

TURNAGAIN ALLEY All of us raised in a city can remember the outlying fields in which we played as youngsters. We remember them better because today they are no longer fields but a part of the solid city itself, where great buildings have replaced the sandlot, and paved streets have cut through the meadows in which we used to lie and watch the drifting of the clouds. Cities are always growing, and we measure this growth, as we see it, on the scale of our own individual lives. What we seldom think of, however, is how the city looked one hundred or two hundred or even three hundred years ago. Business, shopping, and sometimes pleasure take us again and again across the old familiar streets and past the old familiar corners and landmarks. So far as we are concerned, these buildings and these streets and corners have looked like this forever. Of the outskirts we can accurately say, "This was just a field when I was small." But of the inner layer, the core and heart of our city, we like to assure ourselves that "it was always like this."

And what nonsense! The very quality of almost

any street corner you want to name implies the history behind it; a record of growth and change which may tell of great days long since passed, or reflect an almost unbroken development of graciousness and respectability to its present healthy status.

Consider the corner of Tremont Street and Temple Place, the present address of R. H. Stearns. Even in that forgotten novel of sixty years ago called *A Little Upstart* a character named Gerald could see on one side of the Common "the never-ending, jingling, jostling traffic of Tremont Street." I wonder what Gerald would make of it today! But long before us, and long before Gerald, Tremont Street was called Common Street. There was little traffic then. It was not much more than a country lane; fields and orchards lay to the south of it, below what is now West Street, and the commercial noise of the busy waterfront was very far away. What interests us most of all, however, is that in 1684, in what was known as Carter's Pasture, a gentleman named Hezekiah Usher, first bookseller of the colony, built a house which enjoyed a varied and colorful career under many tenants down to the year 1830. Usher was an eccentric; so much so, that Mrs. Usher and her daughter by a previous marriage up and left him in 1687 and sailed back to England. Fifteen years later he was accused by Susanna Sheldon of Salem, during the witchcraft flurry, of sticking pins into her; but his brother and Waitstill Winthrop, grandson of the first governor, third occupant of the Usher mansion, and possessor of a poetic name, prevented his prosecution. Other tenants and owners of the Usher House included Deacon Jonathan Williams, a wine cooper; the Reverend Roger Price of King's Chapel; and Stephen Greenleaf, remembered as the sheriff who performed his duties at the soldiers' trial for the Boston Massacre. The next owner, one James Swan, who was born in Scotland, seems largely to have enjoyed a life of litigation. John P. Whitwell, who followed him, had an apothecary shop on Newbury (now Washington) Street and was Boston's first vendor of soda water "equal in strength and pungency to the boasted soda water of London or New York." That was in 1811.

John H. Shaffer, the next incumbent, decided "to devote the house and grounds to purposes of entertainment." In 1815 the premises of the Usher place were opened as the Washington Gardens. Concerts were the first offering, but over the next few years we hear of parties, fireworks, equestrian exhibitions, and even of a balloon ascension with the ascender bailing out. A brick amphitheater was built there in 1819, the year before St. Paul's Church was erected on the adjoining estate to the north.

In 1800, if you had looked through a little gate about seventy-five feet to the south of the present site of St. Paul's, you would have looked down a lane called "Turnagain Alley." It did not run all the way through to our Washington Street. In 1830, when the Usher house was removed to South Boston, the estate was cut up into ten lots. Mr. Walter K. Watkins, who has written the delightful history of Tremont Street and Temple Place, estimates that the value of the entire tract of something over an acre came to fifteen thousand dollars. You would add some ciphers to that today.

The fall of the House of Usher gave way almost at once to the rise of the Masonic Temple at the north corner of the Gardens, facing on Tremont Street, adjacent to St. Paul's. The cornerstone was laid in 1830, and the building, among other things, changed the name of Turnagain Alley first to Temple Court and later to Temple Place. As to that, Temple Place was called Autumn Street for a single week in 1865, and Avon Street for all of two months in 1869. The lower end of it is Avon Street today.

Bronson Alcott, grandfather of *Little Women*, had a school room in the Temple in 1834. Two years later Margaret Fuller assisted there as a teacher in French and Latin. Most memorable of all, it was in this building that Emerson lectured in 1837; and to his famous course there succeeded the present Lowell Lectures, founded in 1839, the year in which the school was closed. In 1858 the Masonic Temple was sold to the Federal Government for a courthouse, and the Government in turn disposed of it at auction in 1885. Completely remodeled, it was first occupied by R. H. Stearns and Company in the summer of 'eighty-six and continued as their store until 1908, when it was torn down and the present building erected.

Temples are noble things, and sometimes sacred. But I submit that in this instance a temple robbed a city of queer names of one of its queerest and loveliest. Tremont is a fine badge for a street, and Temple an ordinary one; but Tremont and Turnagain Alley suggest the warmth and overtone of Threadneedle, Carfax, and Break-Neck Steps. Who could forget an address like that?

BRICK AND GRANITE It is obvious that what gives any city its essential outward charm is the range of its architectural distinction. Even the ghost towns of Colorado achieved a certain distinction of this sort, though it may not have pleased the critical eye. And no one not yet a ghost is qualified to speak of the charm that possibly went with it. The older cities in America present, of course, not one but a composite style of architecture—that is, a series of styles which, in the way they blend or fail to blend, succeed in making the outward impression attractive to native and stranger alike. Boston, a city with more than three hundred years of history behind it, is a kind of architectural museum. You will find within the limits of its urban and suburban circles almost everything from the wooden Paul Revere house of 1676 to a sun-decked modern dwelling within the shadow of a sun-flecked New England apple tree. I cannot attempt to trace the full development of Boston architecture from the early seventeenth-century builders of houses like that vanished one of Cotton Mather to the John Hancock skyscraper soaring toward an uncompleted height not far from Copley Square. But Boston is so rich in exciting examples of architecture in the varying styles of the various periods of its history that it is sufficient to my purpose to speak of a few buildings which give it architectural character, and of some of the many architects whose names are remembered or forgotten.

"No matter what the period, the style was fixed then, as now, purely and simply by fashion and nothing else. No distinctively American style was ever really in evidence, and quite possibly none ever will be, for as long as people lack definite religious and political convictions, and particularly as long as a lagging philosophy fails to catch up with runaway science, just so long will Americans be unable to create a style of their own as distinctive as the Grecian or the Gothic." This opinion I take verbatim from a remarkable little book of one hundred pages called *Boston after Bulfinch,* by Walter H. Kilham—a book which every Bostonian who cares about his city should find time to read. Mr. Kilham laments the fact that an architect's name is not likely to live long after his death. This is a sad truth indeed. Christopher Wren is the one architect of London that the world at large remembers; Bulfinch is *the* architect of Boston, at least so far as the general public goes. Even one with so assertive a name as Ammi Young is forgotten, though his Greek Revival Custom House has since soared to a height unchallenged till this present moment of renewed and noisy building activity.

Charles Bulfinch, who died just before the middle of the last century, was Boston's first native-born professional architect. He built in what we call the

Federalist style. He was practically self-taught in his profession, and he came into his own at a remarkably early age. When he was but twenty-four years old he appears to have submitted a plan for the building of a State House in Boston. And the State House which he did build—whose fresh gold-leaf dome is one of the glories of the city today—was finished in 1795, when he was thirty-two. His buildings and professional influence range geographically from Augusta and Hartford to Washington; but here in Boston we remember him principally for the State House, the porticoed building of the Massachusetts General Hospital, the Otis House on Mt. Vernon Street, three houses on Chestnut Street, the two houses at the corner of Beacon and Park streets, University Hall at Harvard, and the great warehouses on India Wharf. Gone, alas, among other of his works, are the sixteen houses which formed the Crescent on the south side of Franklin Street, penetrated at the center by Arch Street, which in itself was spanned by an arch. This was a disastrous building venture which forced him into bankruptcy; from which he emerged, however, with renewed prestige. Bulfinch was original, daring, and naturally endowed with excellent taste. The circular stairways in his private houses are beautiful enough in themselves to have brought him fame. We must remember that he had but few building materials at his command. Most of his work was in brick, though certain buildings like the Massachusetts General were done in granite. He was apparently a lone worker, had no students, and drew his own plans in his own house. Largely forgotten by most of us is the fact that he served his native town for nearly twenty years as chairman of the selectmen, an office equivalent to that of mayor.

In the Federalist period also was Asher Benjamin, who died but a year after Bulfinch. Some of his early work is up at Windsor, Vermont, but in Boston he built the brick church at Charles and Mt. Vernon streets in 1807, and two years later the impressive West Church on Cambridge Street, now a branch of the Public Library. His own house, which he designed, you will find at Number 9 West Cedar Street. To him are attributed also the colonnaded houses at Numbers 54 and 55 Beacon.

Contemporary of Bulfinch and Benjamin, an Englishman named Peter Banner should be remembered as the designer of the Park Street Church. His plans called for an additional lantern section for the steeple which would have raised it another eighteen feet. The committee responsible for the building—one of those committees, apparently, which frequently interferes with art—compelled the truncation, much to his regret and possibly to ours. The other great New England name in the Federalist era of architecture is that of Samuel McIntire, a man younger than Bulfinch and Benjamin, son of a housewright—that early strain of builders or contractors—born in Salem, and a person of great sensitivity. His was not a Boston name, but it is variously and somewhat obscurely linked with the local architecture of his time.

2. The Boston before Bulfinch has left us a handful of notable buildings without which the city would be artistically the poorer. There is Paul Revere's

house in the North End, dating from 1676; the 1712 House at School and Washington streets, once the inviting home of the Old Corner Bookstore, and probably the earliest example of the gambrel-roof type of building in the city. The 1712 House occupies the site of an earlier dwelling in which lived the antinomian Anne Hutchinson. The John Hancock House on Beacon Street, unfortunately razed just before the Civil War, was the most famous of the gambrel type. On Marshall Street, opposite the building into which is mortised the Boston Stone, stands the Ebenezer Hancock House, a fine brick edifice dating from the early part of the eighteenth century. It was originally one of three or four similar structures on Creek Lane, reaching toward Blackstone Street. As late as the last quarter of the nineteenth century, and probably later, this little stretch was known as Hancock's Row. It is just enough out of the way of the eyes of the curious to escape detection.

It seems a pity that the name of the architect of the Old State House, built in 1713, is now lost to us. Few streets anywhere were ever topped by such a commanding façade. The Old North Church, erected ten years after the Old State House, was designed by William Price, who had seen the Christopher Wren churches in London. The architect of the Old South Meeting House on Washington Street, which dates from 1729, was the work of a man with the engaging name of Robert Twelve. It was an Edinburgh portrait painter named John Smibert, early trained for the trade of a house and coach painter, who drew the plans for the famous Faneuil Hall, built in 1742 and presented to Boston by one Peter Faneuil as a town hall and market. Bulfinch enlarged Faneuil Hall in 1805. As to the regilded grasshopper weather vane, about which many Boston visitors annually inquire, Peter Faneuil is said to have borrowed it from the Royal Exchange in London.

The designer of King's Chapel, built in 1749 and surely one of the loveliest buildings to come down to us, was Peter Harrison, an Englishman who became a merchant down in Newport, where he drew the plans for several notable buildings of that town. There was a stone spire intended to crown King's Chapel, but it was never built. The Ionic columns were added at the close of the eighteenth century by Bulfinch. Peter Harrison's other Greater Boston building, and one not generally associated with the little stone church down on Tremont Street, is Christ Church in Cambridge, erected in 1761.

To an austere city of spires, it was Bulfinch who gave the first softening effect with the dome of the State House. Now the Greek revival in architecture, which swept England and America round the turn of the century, crossed the Boston threshold early in the eighteen-hundreds. Boston in 1820 was a growing city of about 43,000 inhabitants. The desire and need to build touched off the powder of sympathy in professional architects of open mind. The Greek Revival with its columns and temple-like buildings created a refreshing departure from the simple brick and stone structures of the Federalist period. St. Paul's on Tremont Street was built in 1819 from the plans of Alexander Parris and Solomon Willard. Willard carved the Ionic capitals. The material of the front of the building is Aquia Greek sandstone from Virginia, the same material employed in the original part of the Capitol in Washington. Parris also designed the Quincy Market and the building which is now the Somerset Club on Beacon Street. In the 1820's he worked with another architect on the dry dock in the Boston Navy Yard and at certain projects in Boston Harbor. He designed various other houses on Beacon, Summer, and Tremont streets, part of the Watertown Arsenal, and several lighthouses. Solomon Willard, who worked with Parris on St. Paul's Church, designed the Bunker Hill Monument—an Egyptian obelisk rising in the midst of the Greek Revival. Its cornerstone was laid by Lafayette in 1825 and the completed monument was dedicated in 1843 by Daniel Webster. Willard was also the author of other important buildings and monuments, among them the Franklin monument and the fence and gate of the Old Granary Burying Ground and the Suffolk County Court House. He designed churches as far away as Bangor, Maine.

Peter Harrison's King's Chapel had been built from boulders of Braintree granite—what Mr. Kilham calls "the greatest stone construction job yet undertaken in Boston, if not in English America." Along in the 1820's the Quincy quarries to the south and the Cape Ann quarries to the north bolstered the Greek Revival by making possible, if not creating, the age of granite. Indeed, granite was the popular building material for Boston until the turn of the present century. Willard himself purchased a quarry in Quincy, and it is somewhat sobering to realize that each of the monolithic columns for the Boston Court House required a Biblical team of sixty-five yoke of oxen and twelve horses to move it from Quincy to Boston.

The Boston Custom House—the work of Ammi Young—was the first-magnitude star in the galaxy of Boston's Greek Revival buildings. The monolithic columns weigh forty-two tons each. It stood, in 1849, facing the waterfront, but its original charm is entirely lost today under the great tower rising above it. A few of its inner columns now form a kind of modern Stonehenge out in Franklin Park.

The old Tremont House, with a Greek façade of white Quincy granite, was the first of many impressive American hotels which appeared during the Greek Revival. It is one of the vanished buildings of Boston which I would like most to have seen. It must have been a noble affair with that great dining room seating two hundred diners at once, its whale-oil lamps, and bathing rooms in the basement. Dickens said of it that "it had more galleries, colonnades, piazzas, and passages than he could remember or the reader believe." It was built by Isaiah Rogers in 1829 and was torn down in true American fashion in the mid-1890's. It was Rogers who designed the Howard Athenaeum, a granite Gothic building off Scollay Square which burned in 1846 and was rebuilt by the same man. A theater where Joseph Jefferson once played, it is known today the world over as the Old Howard. Another example of Rogers's Gothic style is the Unitarian Church in Harvard Square.

Such are the beginnings of the architectural history of our city. It was the Greek Revival period which, as Mr. Kilham has said so charmingly, marks the point at which "Boston shed its provincial attitude, put on its first long pants, and began to assume the airs of a metropolis."

3. At a single stroke in an age of enlightened art the late Grant Wood electrified the anomalous phrase American Gothic. George M. Dexter's 1847 Fitchburg Railroad Station down on Causeway Street was American Gothic with battlemented towers; but if it is recalled today for any reason, it may be simply because Jenny Lind once sang in the hall on the second floor. Granite with a rough, ragged surface was the American Gothic stone. The Fitchburg Railroad Station and various Boston churches in this brief perpendicular period were built of it; though none of them, vanished or surviving, could equal Richard Upjohn's Trinity Church of the New York Wall Street address. Upjohn, however, was Bostonian to the extent that he had worked for five years with Alexander Parris, and Boston may take some vicarious pride in that.

What Mr. Kilham calls "plain American" as an architectural style came in at the middle of the old century, as Siegfried Sassoon described it, just when Donald McKay was building his matchless clippers. Plain American, you may assume, was not architecturally in a class with clipper-ship style, but a few examples may be noted. The old American House on Hanover Street is one. The Boston Museum, which stood where the Kimball building now stands at Number 18 Tremont Street, is another. The Boston Museum sheltered a waxworks and a natural history collection, but it is chiefly remembered as the impressive theatre where Modjeska, Fanny Davenport, Edwin Booth, and others of the great appeared; to which Copey (Professor Charles Townsend Copeland of Harvard) repaired as a young dramatic critic for the Boston *Post;* for which I. M. Gaugengigl, the Boston artist and etcher, painted a curtain; in which *Pinafore* was given for the first time in America in 1878. Hammatt Billings was the Museum architect. In 1847, ten years after Thoreau graduated from Harvard College, the Boston Athenaeum was built from the plans of Edward C. Cabot. The exterior of this building for me does not match the interior, but the Granary Burying Ground wall is ingeniously niched at the base to favor the graves. Furthermore, the front of the Athenaeum rises far enough back from the sidewalk so that one may view it as one may not view the complicated façade of Tremont Temple on Tremont Street.

Grecian motifs (especially inside) appear to have accented Boston's famous hotels of the carefree day just gone, but Mr. Kilham assigns them all to the American period. They were contemporary with the fabulous shipping era, and I wish I had known more than two of them. I wish I had not been on a diet of milk and flying cereal when most of them were in their ripest years. I wish I had attended the thirtieth anniversary of the Papyrus Club, for example, held at the Revere House on February 14, 1903. The menu was printed in a little paper-bound volume of the members' light verse, which is part of my light verse collection. The prettiest verse of all was the menu itself: Oyster Cocktail, Green Turtle Soup (Sherry), Smelts (Haut Sauterne), Saddle of Mutton (Champagne); Punch, Grouse, Salad, Dessert, Coffee, Cigars, in that order. Simple enough for that day—or night! It was probably well, in passing from mutton to grouse, to have an interval of punch. The postscript to the menu was written by one of Papyrus's past presidents, Judge Robert Grant: "There are certain risks in this world that a man has to take." Arthur Macy, another past president, was more prophetic:

> Gone are those golden nights
> Of innocent, bohemian delights.

The names of those hotels are all still magic: the Tremont House, Revere House, Quincy House (somewhat earlier), American House, Adams House, Parker House, Young's Hotel, and the United States Hotel. The Parker House, alone and in fair replica of the original, is the sole triumphant survivor. Some of the flavor of these vanished hotels is still preserved in Maine in the Augusta House and in the Bangor House which Isaiah Rogers designed.

In the 1850's Connecticut Valley brown sandstone and Roxbury conglomerate began to compete in Boston with brick and granite as primary building materials. Roxbury conglomerate was to show its strength anew, and you may see its durable handwork in the walls of some Back Bay churches, but I do not admire this stone even theologically, and prefer it anchored in its natural ledges. Brown sandstone is another matter, and the then expanding Back Bay and the momentary flowering of the much more beautiful South End pleasantly reflect its use.

If one observes a suspicion of Paris out along the unappreciated Fenway, Paris is actually at our doorstep in the Tuileries-touched City Hall of 1862 on School Street. A Puritan city with a Democratic mayor in a French building looking down at the figure of Poor Richard on a street with an Anglo-Saxon-Latin-Greek name! Gridley J. F. Bryant, who designed City Hall in association with Arthur Gilman, was the son of the man who designed the pioneer granite railway at Quincy. He was another young architect trained under Parris. He enjoyed a wide practice, built many local buildings, including the satisfying State Street Block which Mr. Kilham praises highly as does this layman, and apparently exercised a wide and beneficent influence for plain good taste. Mr. Kilham reports a rumor (undoubtedly familiar to architects) that a French draughtsman or associate in Bryant's office introduced the mansard roof to Boston. Bryant built the classic Charles Street jail which I greatly admire. I should admire it more if the rumored mansard-roof man had been committed to it for his sin.

The old Post Office, which Mr. Kilham reminds us was called by Bill Nye "an inconvenient stone quarry," joined with the Equitable Life building in constituting for him the glory of Devonshire Street. This reminds me that Boston at the middle of the present century is served by two of the shabbiest and least attractive railroad stations in the land. Think, in contrast, of the Grand Central and of the impressive terminals in Kansas City, Cincinnati, and Cleveland. Boston once had several notable stations with national reputations based in part on such factors as first-class restaurants, comfortable waiting rooms with curious luxuries like fireplaces and rocking chairs and "refined wash rooms." The old Boston and Providence Railroad Station in Park Square was a handsome Victorian Gothic affair "built to serve the business of a railroad only forty-four miles long." It boasted an 850-foot train shed which remained for a while the largest in the world. The old *Transcript* said of it that it "was the most Bostonian railway station that ever was erected or ever will be erected," and that it had "a brief but lovely life." No one will say that when the present Park Square bus terminals make way for the rocket depots of tomorrow.

4. Copley Square, many Bostonians will agree, has the most dramatic architectural display in the city. They will not say much about the strays and stragglers along the north boundary of Boylston Street, but who thinks of that? A certain air of antiquity, largely supplied by the massive and handsome Renaissance Library, dispels the fleeting notion that Copley Square (which is not a square at all) did not come out of a box in one piece. However, it did not. The New Old South Church (what other city could have so old a *New* Old South?) was built in 1877. The style is North Italian Gothic—the Ruskin influence. The year before, on the site of the present Copley Plaza Hotel, rose the Museum of Fine Arts: same style, flowery to flamboyant. H. H. Richardson, born in 1838, leaped to fame with his design of Trinity Church, which is in the form of a Greek cross. Copley Square was thus famous before the hotel, which tourists currently regard as the controlling unit, came into being. Richardson's Baptist Church on Clarendon Street —the church of the Holy Beanblowers—was his first Boston commission, and to my way of thinking is one of the genuine flowers in the Boston crown. The lovely frieze on the tower with the trumpeting angels is the work of Frédéric-Auguste Bartholdi, who designed the Statue of Liberty. Richardson died at forty-eight years of age, but he left a varied and imposing legacy to the city, and even to the

suburbs like Auburndale and Brighton, through his series of railroad stations. Austin Hall at Harvard is his work—"Certainly one of Richardson's greatest works," says Henry Russell Hitchcock, Jr.—and I personally dislike it. But Sever Hall in the Yard I do like for all its gloomy mass and dusty interior. The detail of its windows is important for some architectural reason—I forget what--but surely not for ventilation. Richardson in some respects seemed to write *finis* to the architectural sweep that began with Bulfinch. Many buildings have appeared on the Boston skyline since his day; indeed the city seems to be entering a new phase of building activity. But since the death of Richardson there has appeared but one building of magnificent proportions and that is the Public Library completed in 1895 from the designs of Charles Follen McKim of the New York firm of McKim, Mead and White. One may regret the disappearance of the dark red Italian tiles of the original roof, and one may wonder what the inexorable increase of books will do to this work of art should an addition one day seem unavoidable. Meanwhile, like St. Paul's in London, it continues to dominate the Square that bears an artist's name and its immortality seems assured. The curious stir over the gay little figure which Frederick MacMonnies designed for the fountain in the Library courtyard is all but forgotten. Silly and stupid criticism which denied this public building one needed note of gaiety gave New York a good deal of amusement as well as the object of the controversy. In the courtyard of the Boston Museum of Fine Arts there is now a replica of the charming MacMonnies Bacchante. Few people see it there. If someone could get round to making the proposal, a transfer might still be arranged and an old wrong at least partially righted. Why not?

Standing off from the Hill and the Back Bay at a vantage point across the Charles, Boston sometimes reminds me of an old asparagus bed with the stalks of the few fresh skyscrapers and the Custom House tower in lonely silhouette. Design kept the tall white Price building in Quebec from ruining its medieval profile. I am not certain that blueprints from here out will keep the Boston skyline comprehensive and intact. The latest additions in the Copley Square region have robbed some lower neighbors of needed light, but they are each interesting for opposite, not mutual, reasons. Let me sound a clerihew:

Ralph Adams Cram
One morning said damn,
And designed the Urn Burial
For a concern actuarial.

PHILIP HALE AND H. T. P. In the field of dramatic and musical criticism, Boston was fortunate in having in the nineties and throughout the first third of the present century two versatile critics whose reputation extended far beyond the periphery of the circle in which and for which they wrote. One was Philip Hale, whose name rhymes with the college from which he graduated; the other was H. T. Parker, an alumnus of Harvard. They were thirteen years apart in age—Mr. Hale being the older. Mr. Hale's name is identified with the Boston *Herald,* Mr. Parker's with the old Boston *Transcript.* At the end of their separate lives they formed a sort of binary—twin stars, as the astronomers would say—but with this difference: they worked as far apart, as remote from each other, as if one were in San Francisco and the other in Boston. On the other hand, they were both referred to by the close of the 1920's as the deans of their profession. Philip Hale was the scholar, H. T. Parker the artist. The one wrote as a scholar-critic would write, the other as a creator-critic. Nothing could be better for a cultivated community than to have two such giants appearing daily in the public press. They differed sharply in views and approaches, though not personally with each other. Philip Hale wrote directly and simply; H. T. Parker wrote voluminously and with a style at once complex and—to many—enormously provoking. The thing was that they both knew what they were talking about. They had something to say: they were vigorous in opinions, firm in background, taste, and cultivation. Boston was richer for their presence, and it is no secret that the dramatic and music profession, as well as the ballet, looked to Boston as to New York and London for critical authority in these three fields. Philip Hale, in his versatility, conducted in the *Herald* a daily column of prose and verse in extract under the title of "As the World Wags," the like of which has not since been seen in any American newspaper. And it was he who wrote, from 1901 until his death in 1934, the learned, fascinating, and digressive program notes for the Boston Symphony concerts.

There was a further difference between the two. Philip Hale was not, in any general sense, what the world likes to call in quotations a "character." H. T. P. most definitely was. Parker also died in 1934. Two years before his death, I had occasion to write something about him. In reading over now what I said then, it is crystal clear to me that I ought not to speak of him in the past tense, or shift to yesterday for a man who was so intensely of the journalist's today. It took a fast shutter and a stop watch to make anything at all for the record. In some ways he was enough ahead of his time to demand the future tense.

Dickens would have loved him. He was a bachelor, of course; and he lived, in the last period of his life, in the Hotel Vendome—respectable, Puritan, and the excellent haven of neo-elderly ladies. Because he created it, he was constantly in his element. A small, fierce-eyed individual, impatient, impetuous, and abrupt, he was almost always alone. He did not visibly desire friendship. Sometimes in the theater lobby, to which, between the acts, he would repair with the agility of a cat, you might find him conversing in sudden excitement with another member of his staff, or possibly with the manager. But you may be sure, in the second case, that it was the manager who began. He had an infallible ability of making his first words sound curiously like his last. At the Symphony his seat was always to the right and above the orchestra, where he perched, a small and bitter gargoyle above the Brahmin Sea. Since late in the nineteenth century, when he first became an harmonious initial, he was known, read, feared, damned, and praised as H. T. P. In Boston, these were letters insidious as G. B. S., and many a New York manager's complexion suddenly paled or freshened at what was abundantly said in type above them.

He lived and wrote with a fearful urge, though he always claimed to do things leisurely. If Theodore Roosevelt was strenuous, Parker was devastating. His thirty years on the *Transcript* reveal a prodigious amount of work performed with the utmost attention to detail and accuracy. A column and a quarter to a two-column daily review, a daily page to edit, two pages of magazine material for Saturdays, monthly ventures to New York, and a vast amount of consequent reading left him, in the season, time for nothing but more work. Prolific, he belonged with Sainte-Beuve, Mrs. Trollope, and Edgar Wallace. Friends who were interested once calculated that in those thirty years he wrote and printed the equivalent of three hundred full-sized novels, or close to a novel a month. He read with incredible swiftness the essentials of fifteen papers a day. Work was his medicine, and for the loafer he had the artisan's contempt. Although he never rose before eleven, his day was longer than most men's. At two o'clock he arrived at the office, having walked some twenty blocks to get there, climbing three dismal flights of stairs rather than submit to the doubtful mechanics of an old sweet-smelling hydraulic elevator. Three steps down to the "coop," as he called it, a den the size of a good linen closet, which he shared with his assistant, who had been there since nine o'clock. There were never any cheery "good mornings." If he said anything at all it belonged back in the middle of yesterday's conversation. On Mondays he took the most important play or concert for himself; the next went to his assistant; the others —quite often as many as seven—he assigned to what he labeled "various hands." New shows in Boston usually opened on Monday; but he always had concerts, revisits, plays which others had reviewed, and such, to absorb his week. On Thursdays he prepared a Boston letter for the dramatic page of the Sunday *New York Times*. He wrote all notices religiously locked in his room at the Hotel Vendome. All this in longhand, for he never learned or wanted to use a typewriter. He went to bed, on the average, at 4 A.M.

Friday mornings from October until June began a weekly ritual. Roused at eleven-thirty by a messenger with the proofs of his last night's work, he would spend an hour correcting these carefully and penciling, in his atrocious and utterly illegible hand, private remarks to the proofreader such as "You idiots, can nobody down there read English?" Breakfast over at two o'clock, he rushed in a taxi to afternoon symphony, prodding the driver in the back with his cane. At five he was in the *Transcript* office where he glanced at the advance publicity releases, threw them all in the wastebasket, and wrote his own. It is curiously true that Parker knew more about a play coming to Boston than any of the advance agents. If he hadn't actually seen it, he had clipped, read, and digested every New York or London review and printed portions of them in the bargain. At seven, on Fridays, he departed to a monumental dinner in the menumontal style—in the old days at Young's, in the latter at Locke's or the Parker House. Once it was tripe and beefsteak *à cheval;* later, more usually, mock turtle soups, planked steaks, no dessert, no cigars, and black coffee. Back at the office again, he had cleaned up the Saturday magazine pages by twelve-thirty. Supper at one, at the Vendome, of Westphalia ham and Gouda cheese. After that he wrote two columns on the symphony, finishing often as early as six in the morning. He slept till eleven, raced through the Saturday papers, attended an afternoon concert, heard the identical Friday symphony repeated in the evening, retired to his room, and wrote two or three more columns. By then he was really spent and liable to sleep twelve hours. On Sunday afternoon his week's work began.

Every summer he went to Europe, gave up cigarettes, and smoked a pipe. By synthetic accident of foreign clothes, tri-lingual facility, and the Continental manner, he was assumed in London to be a

Frenchman, in Paris a German, and in Berlin an Englishman. But people crossed the street to him, as to a native, for directions and advice. He used to say that once at Cambridge, England, he was mistaken for an Oxford don. He was, perhaps, the most colorful figure in the ample history of Boston arts and letters. And how vigorously did he live up to his creed: "I'd much rather write than talk."

MARBLE INDEX For the stranger—the Bostonian knows it well—I have recommended the view of the Hill in profile from the Cambridge shore at a point just in front of the Massachusetts Institute of Technology, or from anywhere on the Cambridge half of the Harvard Bridge. Let me reverse the order, to invite inspection of the Institute itself. Everyone knows M.I.T. as an imposing cluster of grey limestone buildings just a rod east of the Cambridge abutment of that bridge. It is one of the first and most obvious landmarks about which the visitor inquires. The central building itself, capped by the round dome which houses the library, has the very look of solidity—a geometric figure, if not in Euclidean space. The great open court, the tall windows with the names of the mighty—Aristotle, Galileo, Newton, Darwin, Franklin, and dozens of others—carved above them in homage to science; indeed, the whole multiple façade, as neat as the face of a slide rule, suggests the very mystery and power of natural law. More than any building I have ever gazed upon, this one comes nearer to satisfying Wordsworth's equation for Isaac Newton:

> The marble index of a mind forever
> Voyaging through strange seas of thought, alone.

It is, after all, something worthy of remark for an institution to suggest so vividly in its architecture the functional quality of its being. Undoubtedly the most famous technological institute in the land, M.I.T. trains for life and for citizenship as well as for a career. But the career, in all its honor and loneliness and reward, is written on its pylons and across its doors.

It was little more than one hundred years ago, in 1846, that William Barton Rogers, a Virginia geologist and natural scientist, and professor in the University of Virginia, outlined a plan for a school of practical science—a technological institution of col lege grade. Fate deemed it for him a year of indecision, for not until 1861 did the Commonwealth grant a charter for a school of science to be known as the Massachusetts Institute of Technology. Civil War brought further delay; and finally, in February 1865, the Tech—as it used to be called—first opened its doors. It was not a spectacular opening. Six professors sat down with fifteen students in some rented rooms in a private house in Rowe Place and in rooms of the Mercantile Library Association in Summer Street, buildings destroyed in the great fire of 1872. Professor Rogers became the Institute's first president, and his own words in the original charter set the course from his day to this: that the Massachusetts Institute of Technology was to aid "by suitable means the advancement, development, and practical application of science in connection with arts, agriculture, manufactures, and commerce." Equally important, he further stipulated that the Institute should serve "in some respects a higher purpose by leading the thoughts of the student into those wide and elevated regions of reflection to which the study of Nature's laws never fails to conduct the mind." Those were mighty original words in 1861.

One must remember that in the early days of Technology the liberal arts college was then at the very highest level in our educational system. The university, in the inclusive modern sense, had not yet come to full flower. Science was tolerated in the liberal arts curriculum largely as a discipline secondary to the classics. Perhaps the current training in medicine was the nearest approach to a general training in science which American professional schools could offer. As Karl T. Compton, distinguished president of M.I.T. today, has observed: "It is a source of satisfaction to the administration of M.I.T. that, from the time of its founding, it has avoided the mistake of giving exclusive attention to the materialistic and utilitarian side of the technical training it offers." This attitude is implicit in the new four-year integrated humanities curriculum adopted by the Institute.

Today M.I.T. offers training essentially in the fields of architecture, engineering, science, and business management, though it is principally in engineering and science that it has made its greatest name. In attaining its present position it has, to quote President Compton again, "constantly kept before it three objectives—the education of men, the advancement of knowledge, and the rendering of service to industry and the nation." Indeed, it has pioneered in many fields and scientific practices. It pioneered in establishing professional courses or separate departments in electrical, chemical, and aeronautical engineering, and in naval architecture. It has likewise played a major role in general demonstration of the value of laboratory methods of instruction and in giving economics a strong and prescribed place in the national educational program.

Physically the Institute has so expanded that now those original six professors are balanced by a faculty of seven hundred, and the fifteen students have increased to more than five thousand, meeting together not in a few rented rooms but in a huge plant of laboratories and machinery, cyclotron and wind tunnel, radar and electronic calculator—with all the beautiful complication and simplicity of mind and metal, atom and erg.

All this is on a national scale, of course, for only thirty per cent of M.I.T. students are currently from New England; the other seventy per cent are representative of the rest of the country and the world. A veteran herself of several wars, today the Institute joins her sister colleges and universities in educating the veterans of the last one. Like Harvard, whose laboratory facilities are surprisingly large, she contributed enormously to the war effort in research on war projects, in training, and in many other ways.

Every institution honors its founder or founders. M.I.T. handsomely commemorated her first president with the completion a few years ago of the new Rogers Building at the western end of the main educational group. Many honorable names stand on the role of alumni of this Institute: George W.

Pickering, Hiram Percy Maxim, Pierre S. duPont, Alfred P. Sloane, Jr., Gerard Swope, Edwin S. Webster, Godfrey L. Cabot. One may remember, too, for the roundness of the picture, that Charles H. Woodbury, the artist, was an Institute man. Which reminds me that just beyond the portals of that great house of science, on the most exposed surface of the temperamental Charles, a man may lean against the bridge and watch the unpowered twelve- and twenty-footers of the colorful M.I.T. yachting fleet in action. The Institute is proud of the fact that it has the largest such fleet in the world. Here is science in its simplest and most relaxing terms: a study in triangles, perhaps, and again a symbol in the Boston equation.

My Book and Heart
Shall never part.

BOOK AND HEART Long ago there was published locally a small book of extraordinary and persistent influence. It was called *The New England Primer,* reflecting in a devoutly poetic way the temper and spirit of the age that produced it. I was looking not long ago at my small facsimile copy which carries one swiftly back to the close of the eighteenth century. There is more than a scattering of verse in *The New England Primer.* I seemed to remember, on turning those pages originally printed in Cornhill, that a book only recently published has borrowed its title from one of the most imaginative verses of all. This reminded me in turn that the motto of *A Witness Tree,* by Robert Frost, is taken from the last of the alphabet rhymes, which you can look up, and that Thomas Wentworth Higginson in 1897 found a three-word title for a collection of his essays in another of the same rhymes. What Higginson found was

> My Book and Heart
> Shall never part,

and he called his volume after the first line.

My book and heart shall never part. How true those words were in prophecy for us. Today Boston is the third greatest library center in the United States, exceeded only by the two cities which you would guess: Washington and New York. Indeed, Greater Boston has many more books than hearts, for its various public, academic, and special libraries can show a grand total of some twelve million volumes. The University Library at Harvard, with a collection of five million books, is not only the largest university library in the United States and probably in the world, but the second largest library of *any* sort in the United States. Only the Congressional in Washington exceeds it in size. Harvard's library is actually composed of some eighty separate libraries, many of which are of importance in themselves. For example, the Harvard Law Library, with some 625,000 volumes, contains the finest collection of foreign law to be found anywhere, and it has also at least the second-best library of Anglo-American law. Harvard's Chinese-Japanese Library, of nearly 200,000 volumes, is one of the great collections existing outside of China.

In the center of Boston, at Copley Square, stands the Boston Public Library—the oldest free municipal library supported by taxation in any city of the world. Founded in 1852 and opened to the public two years later, its first quarters were in the Adams School in Mason Street. In 1858 it was moved to its own building in Boylston Street, on the site now occupied by the Colonial Theatre. It was not until 1895 that it once more headed west to the present beautiful Italian Renaissance building designed* by McKim, Mead & White, there through simple ac-

*In 1887.

cretion and the passing of years to become one of the great public libraries of the world. Everyone knows that this is a building of more than books, with its inner courtyard of fountain and trees, cloister and quiet; its dignified and impressive stairway and rooms, and rich architectural detail; the murals by Abbey, Sargent, and Puvis de Chavannes; the statues of varying merit by St. Gaudens, MacMonnies, Daniel Chester French, and Bela Pratt. But not everyone knows or remembers that functionally it serves the citizens of our community through thirty branch libraries, twelve of them housed in buildings owned by the parent library, and through the Kirstein Business Branch in the Kirstein Memorial Building at 20 City Hall Avenue. This business branch alone, the gift of the late Louis E. Kirstein, is used by more than one thousand people a day. Only eight other public libraries in the United States have such a branch. And that is not all. Through an agreement with Harvard University, the Baker Library of the Harvard Graduate School of Business Administration is also a branch of the Boston Public Library system, open to the public for reference. Through a similar agreement, the Boston Medical Library, the third largest library of its kind in the United States, has for many years served as the medical branch of the Boston Public Library. Over in Brighton, the New England Deposit Library, opened in 1942 and the only one of its kind in the country, takes care of the lesser-used books of eleven different institutions in the Boston area, including the Harvard Library.

The individual unacquainted with his public library has missed something in his education. If ever there was visible in one place a reasonable cross section of a nation, it is in a library reading room. People go there to read for information, for inspiration, for recreation, for solace and comfort, out of despair and loneliness, to strengthen belief, to brighten the rusty armor of courage and hope. The very old go there; the children use it. The scholar, musician, artist, teacher, writer, editor; the professional man and the businessman—the fellow far from home—all turn to it every working hour of the day. You can tell the public library that really serves merely by looking into the reading room through the open door. The public libraries in Cleveland and Rochester, for example, are the kind that really serve. Not the shelves alone but the whole building in these two cities invites you in. So with Boston's great library and its satellites.

The list continues. One of the great libraries for the blind, with the finest collection of books, pamphlets, and periodicals about the blind to be found anywhere, is in the Perkins Institute in Watertown. Boston has seen and sponsored many pioneering ventures in the library world. The Boston Mercantile Library, organized in 1820 (a dusty relic today), was the first of its kind in America; the Boston Y.M.C.A. Library, organized in 1851, was the beginning of a long series of them now flourishing under those familiar initials. The Massachusetts Historical Society's library is the oldest in that field. The Massachusetts Horticultural Society's Library, founded in 1829, has the largest and most comprehensive collection of any such organization in America, possibly in the world.

Number 10½ Beacon Street, overlooking the Old Granary Burying Ground—as every literate American knows—is the address of the Boston Athenaeum.

Though not the oldest of stockholders' libraries in America, it is easily the most famous and prosperous of them all. In a life span of one hundred and forty years it has counted but nine librarians. It antedates the Boston Public Library by nearly fifty years and the Museum of Fine Arts by more than sixty. In the beginning it rendered a public service with books and pictures and sculptures, but in these latter years it has become largely the private reader's and scholar's retreat. From 1822 to 1849 you would have found it down on Pearl Street. But it is in the present original and other-worldly building that the Athenaeum has grown in tradition and influence. No other Boston institution has anything like its unique, endearing, and enduring atmosphere. It combines the best elements of the Bodleian, Monticello, the frigate *Constitution*, a greenhouse, and an old New England sitting room. Some of the staff members have worked there all their lives. The only typewriter you are likely to see is a Hammond, c. 1886, still complete right down to the curved wooden cover. Yet there is on every one of its five delightful floors an unobtrusive sense of efficiency gained without noise or speed, punch cards, or placebos. The Athenaeum is a kind of Utopia for books: the high-ceiling rooms, the little balconies, alcoves, nooks, and angles all suggest sanctuary, escape, creature comfort. The reader, the scholar, the browser, the borrower, is king. An anechoic chamber is not more silent. Even the elevator is hallowed; and the prints, pictures, and statuary—from Lafayette to an incredible Little Nell—wear the look of ineffable charm. Accessible shelves and stacks are rich and even great in Colonial and Confederate collections. Here is George Washington's private library; here you will also discover the latest (mark you! the *very* latest) books and that variety of carefully chosen volumes in many fields which makes it possible—as Mr. Walter Muir Whitehill, the present librarian, has well said—to dig a considerable distance in almost any historical, literary, or artistic subject before running out of material.

The visitor to the Athenaeum will be impressed by these many things: but first of all perhaps by the view out the great south sunny windows across immortality to the stream of city life beyond. He will also be impressed by the nineteenth-century economy of Athenaeum tea. He will not forget it, because afternoon tea for the reader and scholar still costs just three cents, with a choice of three plain crackers, or one plain and one sweet.

FRENCH BLUE AND INDIAN RED

One of the best windows of escape in any large city is the window of a good commercial art gallery. The best window shoppers in the world are the people who stop there for a moment, to replenish their souls and travel as by some miracle across an unfamiliar landscape. Sometimes it is simply a portrait—the head of an elderly man or the face of a child—that attracts their fancy, but more often it will be the slant of a hill, sunlight on a field, the blue shadow of a mountain, or the stretch of a narrow valley that captures the beholder with its sense of strange unrest—the old horizons again, looming far in the countries of the mind.

You may remember what Mr. J. B. Priestley, in a rewarding book called *English Journey*, said about all this. He was speaking of a watercolor in the Art Gallery in Birmingham. "They have here," he said, "a little Harvest Scene by De Wint—a tiny wagon or two, then a glorious melting distance of rolling country and sky—that I should dearly like somebody to steal for me. It lit up my morning. All the years between Peter De Wint and myself were annihilated in a flash; he pointed and I saw, he spoke and I heard; and his mood, felt on that autumn day long ago, was mine."

Boston and the public have long enjoyed an affinity in art. A glance at the subject in *King's Dictionary of Boston* published in 1883 tells me that "the cultivation of the fine arts in Boston, notably of painting and sculpture, is extensive and wide-spread, growing and expanding year by year; and the city ranks as an art-center second to none in the country except New York." That was sixty-five years ago.

Now if there is one department of the fine arts in which Boston subsequently became not only a center but a powerful stimulus, it is that of water-color. In some respects this is curious, for water-color is a dynamic, exciting, and fluid medium—not one which you might think would be attractive to artists with the more solid and patient traits of New England character in their background. The fine and delicate school of English watercolorists which Mr. Priestley extolled so pleasantly should prepare us for the Boston counterpart. Somehow it does not; for watercolor, which enjoys the lovely French name of *aquarelle*, is imbued with French vivacity. A good watercolor, say the French, is a well-controlled accident, which again is not descriptive of any phase of the technique of a true New England craft such as the hooking of a rug, the laying of a keel, or the fashioning of a Windsor chair. Nevertheless, if the association between watercolors and Boston is less strong than that between whaling and New Bedford, it is there just the same.

Winslow Homer was born in Boston in 1836, and we sometimes forget that fact. I speak of him here solely as a watercolorist, but his influence in that field alone would lift the art reputation of his chosen community to a very high plane indeed. I venture the opinion that he is still the most vigorous landscape painter in watercolor that this country has produced. He is certainly one of our few American classics. New York did him honor in 1936 with a centennial show, and Boston may be proud that the best and most poetic of his washes in that impressive exhibition were those with a rugged background familiar to us here in the northeast corner. The Boston Museum of Fine Arts and the Fogg Museum of Art at Harvard have between them the great Homer caches in watercolor. So he is not without honor in his own country. He did not found a school; but the names of contemporaries who won their laurels, or part of them at any rate, in the Boston circle were and are very largely individualists in manner and style. John La Farge was born a year earlier than Homer, and they both died in 1910. A Rhode Islander from New York, La Farge had a generous and loyal Boston public. The third exact contemporary of these two was F. Hopkinson Smith. Though born in Baltimore, he also had a finger in the Boston watercolor pie. After these three came painters in the medium like Sargent, Hallowell, Frank W. Benson, Dodge MacKnight, and the late Ralph Gray. Mr. Benson is still painting in watercolor, and with all the old magic of his violet, yellow, and blue harmonies; the stream and marshland, and ducks and other birds, and New England winter for which his pictures are memorable. A younger watercolor crowd in Boston today is active with frequent one-man shows. In it, or connected with it, are such well-known men as Sutton, Whorf, Ripley, Lavalle, Shepler, MacNutt, Gallager, and Jewell. Andrew Wyeth, the extremely talented son of a famous artist father, shows regularly in Boston his sharp or fog-touched papers from the coast of Maine. Eliot O'Hara, founder of that widely patronized watercolor school down East, earned his national popularity for technical brilliance in early Boston shows.

Boston's art galleries, apart from her great museums, have long been the friendly meeting ground between the artist and his public. Homer's extreme popularity in the 1890's and on down into the early part of the century was no play on his part to the gallery—rather, the gallery played to him. Doll & Richards, now on Newbury Street, whose history goes back to 1848, recalls the heyday of the Boston watercolor vogue when a Dodge MacKnight show, opening in the twenties, would find a roomful of people waiting to get the pick of his landscapes as soon as the doors were unlocked. More Bostonians than you might guess own today from two or three to twenty Dodge MacKnights, covering subjects drawn from the Cape to Shelburne, New Hampshire, and out to Bryce Canyon.

Some of the old galleries, like Williams & Everett on Washington Street, are gone; but Vose, the Guild of Boston Artists founded in 1914, the Institute of Contemporary Art (late of Modern Art), Jordan Marsh, the Boston Art Club, and the St. Botolph Club, for example, are all independently active in offering annually to the public a variety of attractive watercolor shows.

Not long ago the St. Botolph Club on Commonwealth Avenue sponsored its first invitation American Watercolor Exhibition, with more than one hundred pictures divided concurrently among the Club, the Doll & Richards, and the Vose Galleries. This was a pioneer venture in the city, and its range extended from reactionary to ultra-modern. It was no insular affair. The Far West was well represented by some brilliant studies of Taos, the Mojave Desert, and Indians. The whole exhibition proved a fresh and exciting demonstration to an interested public that Boston in itself is a living and chromatic study as yet unframed.

PUDDINGSTONE It has often occurred to me that though a little knowledge is perhaps a dangerous thing, a little knowledge of geology is a perpetual source of pleasure to the possessor. Certainly in my own case, travels through America, Canada, and certain parts of Europe have been much more rewarding for the acquaintance, however little, I have with this elemental science. The endless boulder fields of New England, the veined quartz of Oregon, the fossil ferns I used to find in Pennsylvania, the crystal porphyry which you have seen (for example) in the Blue Hills, the folding of mountains such as one may observe in a cut like the Delaware Water Gap, the great and marvelously colored cross sections of timeless rock which spellbind the visitor at the rim of the Grand Canyon, or the simple glacial sand plain which one crosses on the road to Emerson's Concord—all these visible parts of the earth's structure have an interest in themselves. They have more than that: they have a fascination, for they tell us something of our planet that was written millions of years before the dawn of primitive man.

A great city like ours, which appears on the surface to conceal the nature and structure of the earth on which it is built, will actually reveal to the patient explorer certain geologic secrets which the original bare hill or plain or valley would not reveal. Of course, when the older part of a city has become completely covered by brick and stone and pavement, as Manhattan Island would seem to be, only the construction engineer, the subway builder, and the sublithic people who attend to all the urgent piping and wiring problems underground have a chance to examine the local geology. That is particularly true of a city built on a relatively flat or leveled surface. Boston, however, which is not built on a flat or leveled surface except in man-made part, still exhibits at curious points the old scars of excavations and mortising which permit one to see the nature of its substructure. The harbor and all the islands in it, about which I have been talking, have their own story to tell. So have the hills like Corey, Savin, Dorchester Heights, and Beacon. The sinuous quality of the Charles River is in itself an open chapter in Boston's geological history. When you are annoyed for days and weeks on end by the rhythmic driving of great piles on which will rest—but not as soon as you hope—the foundation of another office building, you are being told in a loud and explosive voice something else again about the geology of the city in which you live. Or if some other bit of construction calls for a stick of dynamite, the great book of earth has been turned back to an early chapter which you may quickly read if you care to do so. It does not require more than the most elementary acquaintance with geology to be able to interpret intelligently any good contour map of the Boston area.

Everyone has heard of the Braintree slate, the oldest rocks in Greater Boston, formed in the ocean perhaps half a billion years ago, and now exposed to the eye in the Weymouth, Braintree, Quincy, and Milton region. Slate Island in the harbor is composed of this rock in which are found the fossil remains of the trilobite and other primitive creatures of the sea. There is other slate in Somerville, and slate under Belmont and Watertown. All America knows about Quincy granite, an igneous rock which was liquid once (but not to the quarryman!) and is now the most durable part of many of the older buildings here and in other cities. The first railroad in America, incidentally, was the one built more than a century ago to carry the granite from a quarry in West Quincy to the Neponset River. The granite which built the Bunker Hill Monument came out over this road.

Probably few people think of Boston as having once been volcanic. Californians have seen an active volcano in Mount Lassen, and dead volcanoes are part of the grander part of the Western scenery. Well, according to my underground informant, there is a dead volcano in West Roxbury near the corner of Washington and Grove streets. This area is too built up to show the lava rock today, but there are other such rocks, for example, in Mattapan, not far from where Blue Hill Avenue crosses the railroad. There is a lava island offshore from Green Hill in Nantasket, and other volcanic evidence is aboveground in Atlantic Hill in the same region.

Who has not heard of Roxbury Puddingstone or

Roxbury Conglomerate—a cemented mixture of pebbles and sand created by pre-glacial streams which washed and wore away the rocks of the sea's edge, spread them in layers, mixed them with clay, and continued the process until pressure on the bottom layers hardened them to rock? It is a very famous rock formation. It was created or laid down under the sea, and how it came to lie about the hills of Roxbury, Savin Hill, Dorchester, Milton, and elsewhere is another story. To Dr. Oliver Wendell Holmes it was the legend that mattered, and he reported the legend in an early poem,* and talked about it again in 1860 in *The Professor at the Breakfast Table.* A giant and his wife and three children lived in Dorchester. One election day, when he and other giants were choosing a king—a democratic process evidently—he left home, first being careful to fasten his three children in a pen. But the children cried—or roared—and the giant brought them a plum pudding stuffed with plums. The mother and the children, however, fell to fighting and flung the pudding over the Roxbury hills.

> Giant and mammoth have passed away,
>   For ages have floated by;
> The suet is hard as a marrow-bone,
> And every plum is turned to a stone,
>   But there the puddings lie.

2. If you will imagine a great blanket or an apron spread between Arlington Heights and the Middlesex Fells on the north and the Blue Hills on the south, you will have some idea of what is meant by

**The Dorchester Giant.*

the Boston Basin. If you will imagine, too, that the sea once covered this area and that into it and out beyond to what is now the inner and outer Boston Harbor poured an alluvial flow of gravel, sand, and natural cements plus an occasional layer of lava from the local volcanoes, you will have a better idea of how the familiar rock called Roxbury Puddingstone or Roxbury Conglomerate was formed.

As these layers of forming rock increased over a period of thousands of years, the total weight increased also. It increased to the point where the lower layers began to harden into the Puddingstone as we know it today. The weight increased finally to a point where the great blanket or apron already under the sea suffered a sinking spell of perhaps several hundred feet. The sinking, of course, was accompanied by earthquakes and other natural violence. The solid granite of the Arlington Heights and Middlesex Fells region on the north and the granite of the Blue Hills on the south held fast, if they did not actually rise a little. At these points appeared the fault lines, which is the geological term for the plane along which the sinking of a mass of rock occurs. The San Francisco earthquake was caused by such a slipping of one part of the city's substructure. Pleasant Street in Arlington lies at the foot of the northern wall of Boston Basin, marking the boundary fault.

This sinking of this Basin occurred millions of years before the ice sheet crept down from the north. But even before that frigid invasion, the troubled earth put on still another act. The two granite shoulders at Arlington and Milton were gradually pushed toward each other. The sunken conglomerate, lava, and the older slate rocks undersea in the Basin were buckled into fluid folds, much as if an elephant stepped on a five-gallon can of molassses. As a result, the land in the Boston Basin actually rose from the sea. It was land made out of a mixture of distortion and compression, arches and curves. Some of it created the promontory on which the new city of Boston was one day to be built. Other parts of the upheaval long afterward became islands in the harbor. Then the slow process of erosion began all over again, and rivers and running water commenced the age-old task of leveling. We may look at the hills in Roxbury today as a testimonial to the toughness and surviving power of the conglomerate that bears its name. Other and softer rocks have long since weathered into oblivion, but not this puddingstone structure of pebbles, cobblestones, and natural cement.

The area of the Boston Basin which rose so dramatically out of the sea and gave us the land on which we now live and have our city presented, even a million years ago, a vastly different topographical appearance than it does today. It was not then sufficiently eroded or worn to show the hills and valleys as we see them. We know, for example, on substantial geological evidence, that a million years ago the Merrimack River did not flow from Lowell to Lawrence to Haverhill and thence to the sea at Newburyport. It flowed from Lowell down to the Mystic Lakes, Spy Pond, and Fresh Pond—none of which then existed—crossing the present course of the Charles somewhere between Cambridge and Brighton, finally flowing out to sea below the present South Boston. The Charles itself joined the Merrimack about at Allston, but it was not so crooked as today, and there were no falls in it.

All this will give, perhaps, an idea as to how the Boston area looked when New England and North America entered the Ice Age. There was again an uplift of the land, and the climate grew colder. The great ice sheet descended on us from a center in northern Quebec and the Labrador. It took thousands of years to reach Boston; it remained here thousands of years before it melted away. In that time it covered the tops of the White Mountains. It scoured the land: smoothing and polishing ledges, pushing fragments of rock and boulders ahead of it, leveling hills and gouging depressions. It passed on down the eastern coast as far as Long Island, New York.

At length, after many centuries, when the ice sheet began to melt in a gradually warming climate, it left great deposits of gravelly rock and debris which—according to their appearance as hills, mounds, ridges, and so on—are called drumlins, eskers, moraines, and by other interesting names. The depressions which the ice had scooped out later became lakes and ponds. Fresh Pond in Cambridge, for example, was cut out by the advancing ice which left a kettle hole. So were Jamaica Pond and Spy Pond. Harvard College stands on a very low drumlin. Milton Hill and Brush Hill, on the other hand, are large drumlins. Corey Hill, some two hundred and sixty feet in height, is a very large drumlin. Fisher Hill and Walnut Hill are others. Out in the Arnold Arboretum, Bussey Hill is a drumlin; so is Peters Hill.

As to what we call old Boston itself: Beacon Hill, its largest ice-made drumlin, has long since lost through the work of man the triple nobs which gave it the title of Tri-Mountain, whence derives our present name of Tremont. Copps Hill is a small drumlin. Fort Hill is another, but it has been leveled by Bostonians into a square. South Boston stands on three drumlins; and Dorchester Heights is indeed an exceedingly large one.

One of the interesting things which the retreating ice sheet did was to litter a great part of New England with boulders, which made it difficult for the early settlers to clear their planting acres, though it gave them in the end the most durable of meandering walls, and literature a lasting poem. The ice sheet played some odd tricks, too, and left evidence of its playfulness in a few great balanced boulders—one above the other, such as you may see in the Middlesex Fells.

When the ice sheet melted locally, the west side of the Boston part of it went first. A tongue of ice crossed the mouth of what is now the harbor and held back the water from the melting ice just to the west. This formed what in the geologies is known as glacial Lake Shawmut. The surface of this lake, the scientists tells us, was some seventy feet above the present sea level. It covered Boston, the Back Bay, South and East Boston, Cambridge, Revere, and much of the immediate outlying districts.

After the tongue of ice to the east in turn had melted and glacial Lake Shawmut had drained away, the Merrimack River, encountering vast quantities of glacial drift and debris, altered its course, flowing to the east as it does today. The old Middlesex Canal and the Boston and Lowell Railroad were built along the original course of the lower Merrimack Valley. The Charles River, seeking new access to the sea, twisted and turned in its effort to get round the various blocks of glacial drift. It crossed the now buried valley of the old Merrimack at Allston and made a new channel round the north end of what was once called Boston Neck.

As to Boston Harbor: after the disappearance of the ice, the seaward land gradually sank again and the ocean flowed into the valley, forming bays and making islands of the glacial mounds and slaty hills. Most of Boston's islands (as I have said) are drumlins, just as most of the islands off the coast of Maine are solid rock. Boston Harbor is really what is called a drowned valley—a type of harbor characteristic of the shore all the way down East. When we think of Boston as a seaport city, whence came and fled the once tall ships, we might remember, too, that she was herself largely built *in* the sea, on whose uplifted floor we walk today.

SEAMARKS Down at sea level the good glass picks up the fragments of trees and buildings apparently afloat in Boston's highly complex harbor. These would be the islands or the indications of islands along the easterly perimeter. But few people carry a glass today, and the floating specks are not often in the range of mainshore vision. As a matter of fact, when we stop to think that this harbor as a whole covers an area of about seventy-five square miles, we need not wonder that the average citizen of the town has but the vaguest notion as to its shape, the location of its principal channels, the distribution of its primary islands, and the natural geometry of its two great breakwaters created on the north by Winthrop Beach and Deer Island and on the south by Nantasket Beach and Hull. It might be the Aegean Sea, for all the report of it he could give you.

In San Francisco one may ascend to the Top of the Mark and view in a single sweep the major area of one of America's most fascinating harbors. Islands and bridges come clear in outline and direction, and man's personal compass makes it easy for him to read the whole magnificent blueprint of the Golden Gate. Even Puget Sound lies reasonably visible to the citizen of Seattle, for the great Olympic peninsula is definition enough for one boundary at least. In Boston there is no central lookout. The Bostonian has but few commanding hills to climb, and these few are not located at the crossroads of general traffic. He can see a good deal of his harbor from Dorchester Heights, from the lofty parts of Roxbury, from Blue Hill, from two or three fire towers, from a certain panoramic station on the Chickatawbut Road, from various high buildings in the city such as the new courthouse and the Custom House tower, and so on. But from none of these elevations does one actually look *down,* as in San Francisco: the view is outward and seaward, and geography is never in clear focus. Even a trip down the harbor under sail or power is not enough. You begin to see things from the other side—the cliffs of islands that you thought were flat, the inhabited islands that you thought to be desolate, the unidentified islands that shouldn't be there, and the mystery deepens. Only a plane can put the dots in true proportion and smooth the wrinkles from the blue cloth of the water.

It was no divine accident that laid out Boston Bay and within it fenced off Boston Harbor by a chain of islands strung between the opposing ends of the peninsulas of Winthrop and Nantasket. This was the work of glaciers, and the islands are largely what the geologists call drumlins—a heap of sand and fragmentary rock formed to the shape of an elongated hill by the drifting action of masses of ice. In this case the drumlins thrust their heads above the harbor floor. Altogether there are more than one hundred islands in Boston Harbor and Boston Bay. They range in size from those whose area is reckoned in acres—some of them rising perhaps one hundred feet above the water—down to half-tide rocks and bars. The majority of islands in the bay are nothing but bare rock; but a few are composed of glacial drift. Most of those in the inner harbor, however, are purely glacial. Of these, a number have cliffs deeply etched by centuries of moving water. Here you will find a few boulder reefs visible at low tide to the seaward side of the cliffs, indicating the original extent of the drumlin structure.

The names of the larger islands are familiar; but the names of the lesser ones are remembered chiefly by cartographers and by the inhabitants of the nearest stretch of mainland. Some of the islands have changed names with the centuries—not always for the better. East Boston on a seventeenth-century map is Noddle's Island. Nut Island was also called Hoffs Thumb at one time, and Moon Island had originally the happier designation of Manning's Moone. What is there in man that gives him so often a positive urge to substitute for the poetic place name something common and utterly devoid of imagination? I wish I knew! Old Dr. Nathaniel Shurtleff, author of *A Historical and Topographical Description of Boston,* took a pleasant view of the physical appearance of Boston's islands. Let me quote him quoting *Hamlet* as he goes. "Noddle's Island, or East Boston as it is now called, very much resembles a great polar bear, with its head north and its feet east. Governor's Island has much the form of a ham, and Castle Island looks like a shoulder of pork, both with their shanks at the south.

Apple Island* was probably so named on account of its shape; and Snake Island may be likened to a kidney; Deer Island is very like a whale facing Point Shirley; Thompson's Island, like a very young unfledged chicken; Spectacle Island, like a pair of spectacles; Long Island, like a high-top military boot; Rainsford's Island, like a mink; Moon Island, like a leg of venison; Gallop's, like a leg of mutton; Lovell's, like a dried salt fish; George's, like a fortress, as it is; Peddock's, like a young sea-monster; and Half-Moon, like the new or the old moon, as you view it from the south or the north. The other small Islands resemble pumpkins, grapes, and nuts, as much as anything; hence the names of them."

When you stop to think that one hundred miles down East it is possible in the summer to sail the length of Casco Bay for the sole purpose of viewing the outlying islands, you raise a local question. How pleasant it would be if some enterprising company should offer the public a choice of island cruises in the inner Boston Harbor. The trip to Nantasket is but one (and by no means the best) leg of the voyage. The owners of small boats know this for a fact; but the man ashore—does he realize that he is anchored to the leeward of his own Hesperides?

2. The history of the principal islands of Boston Harbor is in some respects the most interesting frontier history of the town itself. When the seventeenth century was still young, Captain John Smith and Governor John Winthrop successively, if not successfully, made maps of the harbor and these

*There are those who remember the old tree at the very center and crown which formed the stem.

islands. Somewhat later—about 1685—that energetic pirate Thomas Pound charted the area apparently with much greater accuracy. Now, after some two hundred and fifty years, winds, weather, changing tides, and the designing hand of man have altered the faces of many of these bleak or tufted or cultivated seamarks, and in some cases—Bird Island, for instance—have erased them completely. As far as sea travel goes, the islands of Boston Harbor today are but landfalls to the incoming ships; channel markers in fair and dirty weather; a promise of the mainland just ahead. To the outward bound they are the last green symbols of the continent left behind. But they are also islands with their own queer look and special magic, like those in the Mediterranean, or others strung along the inner coast of British Columbia, or splendid and occasional in the northern green and blue of the great St. Lawrence.

The islands have not wanted for historians. Shurtleff, Stark, Sweetser, for example, and Edward Rowe Snow, the most ardent and affectionate fellow of them all, have unearthed and preserved a mass of fact and legend. Mr. Snow has even walked to his islands. Eleven years ago, in the company of a young lad of sixteen, he trudged across the ice in Boston Harbor, touching at each of thirteen of them, a record only approached in 1852 by a Major Van Crowinshield, who crossed on the ice to twelve.

Several of the islands, if one examines their biographies, are interesting historically for military reasons—among them George's Island with Fort Warren; Castle Island, on which Governor Thomas Dudley landed in 1634, deciding in the company of his twenty Puritan companions that it was worth fortifying; and (to a much lesser extent) Governor's Island and even Peddock's. Castle Island, like some others, is today connected with the mainland. It was a grim-looking outpost during the last war—the seaward terminus of the long and impressive palisado which successfully shut out from the eyes of the curious all operations in the channel which opens into waters called President Roads. It is, of course, one of the best vantage points in time of peace from which to view the outer shipping lanes and all the traffic and patient trawler activity on the fringe of a great seaport. Governor's Island is now the final projection of the huge new airport. In comparing a very old map with a recent chart of the United States Coast and Geodetic Survey, the northern edge of the Northern Flats of the long since vanished Bird Island—an area which appears on the early charts to be quite as large as Governor's, and which man and the sea united in destroying—has apparently come to light again synthetically, at least in part. If that is so, then Bird Island is trafficking once more in wings, though of a different feather from those of the gulls and auks and dovekies which likely constituted its earliest tenants.

Castle Island and Fort Warren on George's Island provide a compounded chapter from the history of the Revolution, the Civil War, and other wars. The best Quincy granite and filling granite from Cape Ann made of the latter a handsome citadel more actively involved in the Civil War than any other New England fort. Within its walls were trained hundreds of soldiers for the Union Army, and in its lower reaches were imprisoned more than a thousand captured Confederates. Fort Warren played an active part in the Second World War, and it is interesting to learn, through Mr. Snow's monograph, that the popular legend of the Lady in Black still persists. She had—she still has—ways of haunting which are both curious and original. The most picturesque evidence of her continuing presence, perhaps, is in the account of three soldiers some years ago who, just at the entrance to the Fort, discovered in the fresh-fallen snow five impressions of a girl's shoe. The footsteps apparently came from nowhere and led nowhere.

Most of the islands in the harbor have queer little histories of their own: some of them to do with Indians, some with piracy, as where Captain Kidd is supposed to have visited Nix's Mate Island of the lovely name; some with peaceable pursuits such as farming and the raising of cattle, some with industry of a sort such as Slate Island—one of the few islands in the harbor, incidentally, which is not of glacial origin. Peddock's is still known for its harvest of turkeys. Grape Island was pure Indian ground and has been fairly rich in the yield of artifacts. Two islands are associated with hospitals. One of these was built on Spectacle Island in 1718, and Long Island today contains the Long Island Hospital and almshouse. Deer Island, granted to Boston in 1634 as a game preserve for two pounds a year, is the site of the Deer Island House of Correction, a city institution on city-owned ground. Thompson's Island, which boasts the finest stand of trees on any land in the harbor, was named for David Thompson, who visited it in 1619. It is, of course, renowned for the Farm and Trade School. This is a fine and worthwhile institution, incorporated in 1814, to the trustees of which the island was sold in 1832 for six thousand dollars. There is a kind of detachment and an air of antiquity to nearly all these islands. Many of them, of course, are watchdogs by day and night for vessels small and large, for harbor and coastwise craft, for anything afloat or adrift. Out beyond the concenter of these islands, except such as Green and the Larger Brewsters, Boston Light, the oldest lighthouse in America, still burns from dark to dawn.

THE GREAT FIRE

Boston, as I have just suggested, is not only a seaport town but has been built in considerable part on land taken from the sea or tidal water. Some of her handsomest residences and a great many museums, libraries, churches, schools, colleges, clubs, hotels, auditoriums, parks, and amusement centers are solidly built on man-made ground.

But if water has helped to shape the destiny of Boston, the opposite and more fearful element of fire has also had a hand. Most cities of any importance have been burned devastatingly in part at least once, and sometimes more than once, in the course of their history. We think immediately of London, Edinburgh, Moscow, Chicago, and San Francisco. Many cities have had what are called great fires, still referred to with a capital G: London had hers in 1666, Edinburgh in 1700, New York in 1835, Chicago in 1871, and Boston in 1872. This was not Boston's first fire of size and marked disaster, for she had suffered at least four in the two previous centuries. Cotton Mather, calling Boston "Lost

Town," observed long ago that "never was any town under the cope of Heaven more liable to be laid in ashes, either through the carelessness or the wickedness of them that sleep in it. That such a combustible heap of contiguous houses yet stands, it may be called a standing miracle. It is not because the watchman keeps the city. . . . No, it is from Thy watchful protection, O Thou keeper of Boston, who neither slumbers nor sleeps." But so vividly did nineteenth-century Boston remember her fire of 1872 that you will find it listed in that marvelous compendium, *King's Dictionary of Boston,* published in 1883, not under F for *fire* but under G for *great.*

Though cows are known to have inhabited the Common for at least a couple of centuries, and are popularly supposed to have laid out the more tortuous and picturesque streets of the town, there is unfortunately no legend to claim that a cow kicked over a lantern on the evening of Saturday, November 9, 1872, at the corner of Summer and Kingston streets. The simple drama of it is that fire was suddenly discovered there at that hour and on that date in a wholesale dry-goods house. Now the corner of Summer and Kingston streets is part of an area occupied in the early days of the city by some of the most imposing residences—square brick houses frequently set in the midst of gardens in the manner of the early and still standing houses of such neighboring cities as Salem and Portsmouth. But the rise of the clipper ship, and the resulting preeminence of Boston as a city trading with the world, changed the character of the region into what we call a business district. The stone and brick which went into the new buildings were topped apparently by the homely mansard roof. These roofs, it appears, were made of wood. Hence, there was a congestion of buildings far from fireproof, far from durable, and not too pleasing to the eye. It was the Great Fire of 1872 that leveled this area, or more than sixty-five acres of it. Some seven hundred and sixty-five buildings were destroyed.

When it is remembered that many of these buildings were the warehouses of wool, cotton, paper, and leather merchants, it is easy to imagine how that fire, under even the most favorable circumstances, could still have been a serious fire. The records have it that no alarm was rung until long after one should have been rung. Added to that is the fact that the horses then stabled in the city, including those of the fire department, were down with an epizootic disease. Indeed, so bad was the plight of those dependent on delivery trucks, that days before the disaster porters, clerks, messengers, and stevedores had already been pressed into service to do the work of animals. Although the fire department had taken the precaution to provide extra firemen, it was soon apparent that the city was in for it. Apparatus came from neighboring towns. Visiting firemen sadly discovered, however, that hydrants and the water system in general had not yet been altered from the residential scale to one adequate to protect a business district. There was soon no need to scatter further alarm, for the skies were a visible red for some sixty miles inland.*

*Mr. Walter H. Kilham, the architect, whose excellent *Boston after Bulfinch* (Harvard University Press, 1946) has supplied me with most of the data in a preceding chapter, takes a skeptical view of the epizootic episode: "Most of the sick horses had returned to duty." He also scouts the popular legend that the fire started in a hoop-skirt factory. A manufacturer of hoop skirts, corsets, and bustles occupied the

Fire fighters and volunteers fought without pause, but it was not until late Sunday afternoon that the flames were actually under control. Oxen were used where the horses failed. The fire swept the length of Summer Street, reached out along the lower side of Washington as far as the Old South Church, and crossed Milk Street, somewhat impeded by the new post office then under construction. To the east and north the burning swept to the waterfront and beyond Pearl. State and other streets were saved by the blowing up of buildings. Hottest in temperature, perhaps, was the area touching Milk, Congress, Federal, Devonshire, and Franklin streets. If there was panic, it was apparently quickly checked. People in threatened quarters south of Summer Street sought Boston Common as a safety zone for their goods and chattels. One aspect of the disaster is modern enough: The incoming Sunday trains were packed with visitors from afar, or relatively afar, with nothing better to do than smell the smoke, view the devastation, and observe another's consequent anxiety at first hand.

The major toll of Boston's Great Fire was seventy-five million dollars' worth of property destroyed. It was largely a commercial disaster, for most of the hundreds of buildings burned were crammed with

---

third floor of the fateful building, but the fire (says Mr. Kilham) started in a wooden elevator shaft. He also observes that "hygroscopic moisture in the supposedly fireproof granite fronts expanded and split the stone into fragments, while the cast iron plating of some of the prouder buildings buckled and curled. It was a lesson in fireproof construction learned the hard way." He thinks that "the finest of the new buildings in the burned district, and the highest, was probably that of the Equitable Life Insurance Company on Milk Street (1873–1875), by Arthur Gilman, torn down to make room for the present First National Bank."

raw materials and manufactured products. Bankrupted insurance companies added a melancholy postscript to the story.

But the city took the disaster in her stride. Within a day or two merchants were selling their wares in public dining rooms; relief centers were established; more than three hundred thousand dollars was rapidly raised for the benefit of those who had suffered most. Chicago, the city that Boston had helped through similar troubles a year before, made a generous gesture of aid.

On the bright side, a new and more impressive group of buildings rapidly rose from the rubble. Stricter building laws made safer structures obligatory. It would be a pleasant codicil to add that these buildings ascended in the flowering period of American architecture. Unfortunately they did not. But there is nevertheless today a flavor to downtown Boston and a queer homogeneity in its many cornerstones which set it apart from the business districts of other American cities. Perhaps it is partly because not many of those cornerstones are square. And again it may be simply because they were set, phoenix-like, in the ashes of an earlier and viable city.

CHARLES STREET NUMBER 148

Out along Charles Street on the riverside, just where it empties into the traffic circle created some years ago by the elevated station, you will pass the ghost of a house which once bore the number 148. A garage occupies the site of that house today. None but the older residents of Boston will remember the building of which I speak. Certainly none will remember it prior to 1866 when it bore the number 37. The late Miss Willa Cather, who wrote of it so charmingly in 1936 in a book called *Not under Forty,* indicated in a footnote explaining that cryptic title that she was writing about things which would not likely have much interest for people under forty years of age. How truly she spoke! The jet propulsion pace of the world today leaves but little time for reflection on ways of life which have sadly vanished from the American scene. We still have authors and artists. We have them by the ton; and we all know of many areas in this country where they coagulate to talk shop, admire and criticize and deplore each other's work. But the day of gracious living and the finer amenities of life have so faded from the scene that it is hard to imagine authors and artists of the present age meeting and dining anywhere as they used to meet and dine at Charles Street Number 148.

The Mrs. James T. Fields of whom Miss Cather wrote as one who knew her, and who lived at that famous address, was the widow of James T. Fields of the publishing firm of Ticknor and Fields. In the middle of the last century when Mr. Fields himself was, in middle life, a publisher of international reputation and a widower, he had married Annie Adams, a young girl of nineteen. Mr. Fields died in 1881; Mrs. Fields not until January 1915, surviving him by thirty-four years.

It is of the widowed Mrs. Fields, of course, that Miss Cather speaks at first hand. Her delightful essay is in part composed of paragraphs which she wrote in 1922 in reviewing Mr. Mark A. DeWolfe Howe's *Memories of a Hostess.* Mr. Howe's book, drawn chiefly from the diaries of Mrs. Fields, whose literary executor he had become, is likely to remain one of the endearing volumes in the history of Boston. Its chapters are largely devoted to the married years of the Fieldses, which cover roughly the ascendancy of New England letters. Its pages give the reader, in Mr. Howe's own phrase, not only a chronicle of eminent friendships, but a genial pageant of the great literary and artistic figures of the period, including some famous visitors from overseas.

Let me put it this way. The next time you walk down Charles Street, remember that at Number 148 there once lived a lady of such distinction and charm as made her the center of a circle of men and women whose names are as familiar as the title of your favorite book. Miss Cather said of her that she was flower-like in her own person and that "at eighty she could still entertain new people, new ideas, new forms of art"; but Henry James praised her best by observing that "all her implications were gay." She was a great hostess. Indeed, what other woman could have gathered under one roof, and made at once so effectively and clearly happy, so many different, specialized, and sensitive people? You might remember, too, on your walk, that this lady still lived there not so very many years ago, and that in her early life she had talked with Leigh Hunt about Shelley. That reaches back! It is now more than a century and a quarter since Shelley died.

The names of those who walked down Charles Street just ahead of you, and turned in at that door, included Longfellow, Emerson, Whittier, Hawthorne, Lowell, Sumner, and Oliver Wendell Holmes. They included Thackeray and Matthew Arnold; Charles Dickens and Mrs. Humphry Ward, Bret Harte and Mark Twain; great figures of the stage such as Salvini, Modjeska, Edwin Booth, and Joe Jefferson; painters like Winslow Homer and Sargent. Dickens, who could and did make as much as twenty-five hundred dollars at a single public reading in Boston, was a beloved visitor, much made of, and one who gave apparently a great deal in return. He told ghost stories in that house; and it is merely an incident in the Charles Street legend that he dined with the Fieldses in the simultaneous company of Agassiz, Emerson, Dr. Holmes, and Longfellow, among others.

Edward Lear, alas, the lonely English landscape

painter who wrote the greatest book of persisting nonsense that the world has ever known, never got to Boston or to this house, though he wanted to. One of his pictures hung in the reception room. I wish he had come, simply for the sake of what Mrs. Fields might have written of him in her diary, and on the chance that he might have sung for her his own musical setting of "The Owl and the Pussy-Cat," the notes of which he never wrote down. Out of her great intuition, Mrs. Fields might have had someone rescue them for posterity. I like to think above everything that Miss Sarah Orne Jewett, the novelist from Maine, was often there. After the death of Mr. Fields it was she who became the really intimate friend of Mrs. Fields. Miss Cather has given the world an affectionate yet unsentimental portrait of this friendship.

As to the house itself: it appears to have been not unlike many others of its period in Boston. The reception room and dining room were on the first floor; the long drawing room, running the depth of the house, was on the second. The front windows of this room looked out on Charles Street, the back ones on a deep garden, with the open Charles River beyond the garden wall. As far as woodwork, heavy carpets, and Victorian hangings could make it, I suspect it was a dark interior. Miss Cather remembers Mrs. Fields in widow's lavender, with a scarf of Venetian lace on her hair. Well, she and her world, and even her house—which Henry James used to call her waterside museum—have slipped off into the past. No: *Hoc indictum volo.*

SMOKEFALL Many years ago, when Rudyard Kipling was living in Vermont, he described the way that autumn comes into the Green Mountain country. He did it indirectly by describing the way that summer departs. I often think of the remarkable words that he used. "Summer went out, red-faced and angry, slamming all the doors of the hills behind her." So it is with the recession or the flooding-in of each season up here in the northeast corner. That wave of October sadness which seems today to hang in the air and drift with the rising tide of color is common to us all. It is the law of compensation working against the vividness, the strength and ennobling power of a beautiful October day. Archibald MacLeish said something about it in an early poem called *Immortal Autumn.** Here are the third and fourth stanzas:

But now in autumn with the black and outcast crows
Share we the spacious world the whispering year is gone
There is more room to live now the once secret dawn
Comes late by daylight and the dark unguarded goes

*To my mind one of the handful of flawless lyrics of this or any century.

Between the mutinous brave burning of the leaves
And winter's covering of our hearts with his deep snow
We are alone there are no evening birds we know
The naked moon the tame stars circle at our eaves

All this country feeling is not lost on the city dweller in New England. Certainly it is not lost on the citizens of Boston. I have spoken of how the spring comes into our streets and avenues by way of the green artery along the Charles and on up into the Common and the Public Garden. Summer goes out in the same mysterious yet visible way. The first dried leaves of the elm fall and curl and scatter themselves from the Frog Pond down Commonwealth Avenue and elsewhere over the acres of other parkland and open squares. The grey-blue and the violet and purple of chicory and New England aster, flowering late in the vacant lots and along the Charles embankment, are a sign. The anxious proclamation of the city crow and the flocks of starlings in critical voice in the tops of tall trees are another. The blue flash of an occasional jay foraging on wide window sills and in back yards; the pale and almost transparent cold green which floods the under quarter of the northern sky just above our rooftops—these are part of autumn too. And the days before and particularly after rain, when the great bronze clouds hang low to the west and north, silhouetting the field of church spires of the Boston suburbs and Cambridge in particular—they are indicative of autumn and of nothing else.

Out of the city and into the hills and mountains of the red and yellow pigment and the pine-needle green, or down to the russet cranberry margins of the Cape, stream the Boston cars. This is the second season of tourist outflow, as indeed you may guess from the many inconspicuous advertisements in the newspapers of country inns exploiting this Indian summer of the foliage trade. Back they come—the same cars—filled with late corn and yellow pumpkins and autumn flowers from the most attractive and colorful roadside stands in America.

About this time at the edge of the city, and all through the suburbs from Milton to Melrose, the air is often stained with the acrid smoke of burning leaves. This is the season when city children, who may never have seen a farm, for once enjoy the rustic privilege of shuffling through the unburned drifted piles on lawns and sidewalks and in the gutters. Though the year is dying, the citizen old and young experiences a certain renewal of life. There is an exaltation about driving down almost any Boston street in October which does not occur at any other time of the year. It is almost as though, for one startling moment, the country and city were one. It is no time to be inside a house.

> The sun lies supple on the bricks;
> I walk the fluent street.
> The year is at its ancient tricks:
> How bountiful with hay and ricks,
> How beautiful in leaves that mix,
> How fitful at my feet!

Of course, the store windows are part of it too. Spring may be the season of millinery and other tokens of brightness, but autumn in Boston has a solid tweed-like depth to it—none of your flimsy ruffles, but the display of sensible, deep-dyed materials in keeping with the outer color of the outer landscape. There is one week of October, indeed, when I never fail to recapture a certain boyhood remembrance of the color of Halloween. It comes at the close of evening—usually near a fruit store or a flower shop—just when the office buildings pour their life stream into the streets and people for the moment seem uniformly gay and animate and kindly, and the lights come on with a special brightness and twinkle.

This above all is the time not only to talk about but to walk about Boston. It is never again so tranquil in the early morning, when the sun puts a pale bright film across the rooftops and chimneys over the Basin, softening in deep purple-blue the aristocratic outline of the Hill. It is the one time when the dark of evening flows quietly in, as sadness will sometimes drift in and out of a human face. It is the season of the end of something and the promise of something else that is not winter. I think perhaps above all that it is the season of dignity, when unconsciously we tend to match ourselves against the brilliance of nature; when our city, old and new, seems even better than she is—as though she were suddenly house-cleaned and painted, and aired by all the winds that blow across her from the land and the sea.

AIMLESS The art of walking is as old as man himself. But the phrase suggests a country backdrop, with mountains, valleys, lakes, and rivers. The art of walking in a city implies a precious and special talent. Many have had it and lived to write of their experience. It is one thing to be a man like George Borrow, a walker in the gypsy manner under the sun and sky and stars when the wind is on the heath, and quite another to be a walker like Charles Dickens in the great city of London, often alone and abroad for hours in the solitary dead of night. George Gissing, the lonely author of *The Private Papers of Henry Ryecroft*, was another who, on foot and largely companionless, knew London as well by night as he knew it by day. The late E. V. Lucas, a tireless wanderer in so many cities of Europe, was still another. Christopher Morley, who has loved Philadelphia and New York as intensely as Dickens or Lucas ever loved their London, is the full-time American wanderer who knows no clock and turns his steps in every direction of the compass, inspired simply by the passing show, unburdened by any thought of destination. "I can be as solitary in a city street as ever Thoreau was in Walden," wrote Mr. Morley in one of his most effervescent books—a very youthful one—called *Travels in Philadelphia.*

"From now until the end of time," he said, "no one else will ever see life with my eyes and I mean to make the most of my chance." Not all of us are equipped as Mr. Morley is to see remarkable things in unremarkable places. If one walks along a city street or down a country road with eyes averted, there will not be much to report at the end of the journey, and even less to store away in the higher levels of the mind. To walk well in the city—particularly in Boston—requires all the cunning of a talented woodsman. Come to think of it, I have used that phrase because I half remembered something that Mr. Morley wrote in his Philadelphia travelogue about his friend the mountaineer. "He tracks down distant streets and leafy glimpses," says Mr. Morley, "with all the grim passion of a Kentucky scout on the trail of a varmint . . . No old house, no picturesque corner or elbow alley escapes his penetrant gaze. He has secret trails and caches scattered through the great forests of Philadelphia, known to none but himself." That is the way that the lover of any great city goes about his business. He is not guided so much by streets and avenues and byways as he is by the secret blazes on the corners of buildings, the signs and lettering over stores, the look of doorways, iron fences, and railings, the gleam of old brass name plates, the intrusion of trees, the openings to the sky, the way the sunlight falls on brick, the way the shadows lie on stone, the remembrance of landmarks gone, the smells and savors, and the growing acquaintance with what is fresh and new.

Being a city of incredibly narrow and winding streets, where the elevated runs underground, Boston in its oldest areas is to some extent a city of compulsory walking. By this I mean that her citizens, who otherwise might drive or ride in taxicabs on the short hauls, are more or less obliged to depend on their own two legs to get them from place to place. This, of course, does not describe the kind of walking that I mean. The mass appearance of humanity on and all over Washington Street every noon, and every fair and every rainy Saturday, is a curious sight to the stranger, but not especially inspiring to the man who is out for a stroll. The great crowds that pour from the subway exits opening in the hopeful direction of Braves Field and Fenway Park by day, and now by night as well, even though on pleasure bent, hurry by completely unconscious of the region through which they are passing.

To walk in Boston, which was made for walking, one must choose the right streets at the right time. A great many of the older generation of Bostonians, and some of the new, take particular pleasure in walking to work in the morning. If one lives within a mile or two of the office, and if that office is situated in the downtown part of the city, it is difficult to think of any route from any direction that does not pass through block after block which are interesting for this reason or that. As one puts the Common behind and sniffs the first salt air from the harbor, there is surely some unconscious delight in the rediscovery that no street leads directly to any place or any thing. In the course of but half a mile, any number of turns and angles have to be negotiated, any number of new vistas suddenly open to the watchful eye, any number of fragments of the city's history flit ghost-like through the mind. Not all of us can walk to work, but those who can and do so enjoy a peculiar experience that is not without value for the balance of the day. It is one of the factors that has kept the actual as well as the legendary Bostonian male the lean and spare figure that he always turns out to be in the histories and frequently appears to be in everyday life.

So diverse are the thousand possible walks in Boston, from Atlantic Avenue to the Common, from Charles Street to the Fenway, from Dover Street to Castle Island, from Louisburg Square to Bunker Hill, from the Massachusetts Historical Society to the Arboretum, from Bay State Road to Soldiers Field—so diverse are they from anywhere to everywhere, that hardly any two people will have identical favorites. As for me, I choose and long have chosen the streets that turn the most and the directions that include some realistic climbing. That beautiful curve at the river end of Mt. Vernon Street, the glimpse of the lion's-head watering trough and stable yard and the granite carriage track and wheel rest set in cobblestones at Number 85, the descent past Copps Hill into Atlantic Avenue and the gleam and smell of ship chandlers' row, the look of Beacon Hill from the parade ground on the Common, the approach to the old city along the Basin from the Cambridge side, the nobility of the elms down Commonwealth Avenue, the seldom-remarked sight of the Park Street Church up the Columbus Avenue axis—the many forgotten and lovely squares—these are some of the images which the walker in Boston remembers more clearly than the man behind the wheel. You will note that I have called him a walker and not a pedestrian. A pedestrian is a man in danger of his life; a walker is a man in possession of his soul.

OUT OF PRINT Boston, like every lovable old city in the world, has been the subject of a great number of books. She has been evaluated in terms of praise and disparagement, delight and dissatisfaction. She was always a city for the pencil and brush. She ought to have had, in their day, the full Henry James and Joseph Pennell treatment, or the full William Dean Howells and Pennell treatment, though somehow she never did. Oxford and London, Edinburgh, Paris, and Rome have each been preserved for posterity in excellent books full of agreeable pictures. London and Edinburgh have been doubly fortunate in the perambulations of two members of the literary and artistic Bone family. Everybody knows *The London Perambulator,* that lovely book of text and etchings which intermingle as naturally as London street cries and the voice of Big Ben. Not every city can have such luck.

But Boston has her own shelf of books—a shelf that is growing every day. Indeed, within the last two years five or six distinguished—or at least distinguishable—volumes have been added, and there are undoubtedly many more to come. I sometimes think that a personal collection of books about the city in which one lives, or the city which one remembers and loves, is something toward which every intelligent citizen might aim. It gives latitude and longitude to one's library. It is an easy collection to begin and never to finish. Books about cities have a way of slipping off into the darker recesses of the secondhand shops, with an appropriate markdown on the inside of the cover. Some of them, of course, enjoy a perennial sale. This is when they possess a certain guidebook and companion value. The stranger from Dallas, Dubuque, and Buffalo is always in the market for the latest word on Faneuil Hall and the Old State House—as if that word had not been printed long ago in the forgotten chapters of our history. No one realizes better than myself how many good things have been said about Boston in the books that were available to Henry Adams and Josiah Quincy and men before them. The chapters preceding this one could not even have been attempted without the chain reaction of book on book, from Sewall down to Esther Forbes. My debt to many authors and many pages exceeds my ability to catalogue. I can do no more than mention some of the principal volumes which have delighted and helped me, to which I trust I have done no great disservice, and without which I should have been myself a helpless perambulator.

My steadiest companion has been a book called *Boston: The Place and the People,* by M. A. DeWolfe Howe. It was published in 1903, and it ought to be brought up to date—that is, if you can ever bring anything up to date in an ever-expanding city.

My copy is characteristic of the rich old days of publishing—fine paper, solid binding, with plenty of gold leaf on the Bulfinch dome of the State House well reflected on its amber cover. Indeed, that little crowning semicircle of gold is just as bright today as the new foil shining in the sun at the top of Beacon Street. Mr. Howe's book is illustrated in the good, grand style by the late Louis A. Holman and others. The author wrote intimately of the city of his adoption, with that affectionate respect for detail and excursion which bears small relation to much of the staccato writing of today. If you would choose a cornerstone for your Boston shelf, it is *Boston: The Place and the People,* written in a day when the public bought books not on a book-club recommendation, but because it wanted to own them and keep them and refer to them. About a year ago Mr. Howe and Samuel Chamberlain, the artist, collaborated to produce a smaller volume, called *Boston Landmarks,* full of excellent photographs, which strikes me as a kind of corollary to the earlier book. It throws new light on old places and substantiates the legend that Boston is one of the timeless capitals of the world.

For curiosity and relic value, I turn to my copy of *King's Dictionary of Boston,* by Edwin M. Bacon, published in Cambridge in 1883. In its more than five hundred pages of eye-splitting type you may read about things come and gone, the familiar and forgotten: about the old Globe Theatre and the Jeffries Fund and the Mystic Water-Works and the Emancipation Group and the electric light. "In very recent years this light has been quite extensively introduced into the city. In 1880 the privilege of lighting Scollay Square was secured by the Brush

Electric Light Company . . . Later other companies obtained a foothold in the city." Think of that now: the *privilege* of lighting Scollay Square! As I have already observed: it is in this enchanting volume that the great fire of 1872 is listed under *Great* and not under *Fire.* Should you want to look up Woman Suffrage, you will find it under *Isms.* In it, too, you will happily come across a full-page advertisement of a Flexible and Reversible Electro-Magnetic Flesh-Brush made of vegetable fiber; a full-page advertisement of Ober's French Restaurant at Number 4 Winter Place, before Mr. Locke joined forces at Number 3; another of the New England Mutual Life Insurance Company when its total surplus was a little over two and one half million dollars; and a reference to Lake Winnepesaukee when it was still spelled *Winnipiseogee,* as it ought to be now.

Drake's *Old Landmarks and Historic Personages of Boston,* profusely illustrated and published in 1873, is a mine of miscellaneous information. For various reasons I am variously indebted to such other books as Mary Farwell Ayer's *Early Days on Boston Common* and Mr. Howe's monograph on the same remarkable tract of land. Anyone interested in the Public Garden should turn to the 1850 Report of the Joint Committee on Public Lands in relation to the Public Garden. *The Islands of Boston Harbor* by Edward Rowe Snow is a generous *omnium gatherum* of island data and folklore. As to the Charles River, Arthur Bernon Tourtellot's volume called simply *The Charles,* in the well-known Rivers of America series, is worth the reading, though it fails to satisfy me as to fish and fishing and tends to stray along the shore instead of keeping to the channel. A good geological primer is Irving B. Crosby's *Boston through the Ages. Days and Ways in Old Boston,* edited by William S. Rossiter, and published by R. H. Stearns and Company in 1915, is a charming item to own, if you can find a copy. Of course, Samuel E. Morison's *Builders of the Bay Colony* and his *Maritime History of Massachusetts* are required reading; so is Esther Forbes's remarkable book called *Paul Revere and the World He Lived In.* As to that, Miss Forbes's subsequent title, *The Boston Book,* beautifully illustrated with the photographs of Arthur Griffin, shows the old city—and lovely corners of it; for Miss Forbes can write affectionately and breezily, and Mr. Griffin has opened his shutter with full advantage taken of the vagaries of Boston weather.

On the entertaining and satiric side, do not overlook Lucius Beebe's *Boston and the Boston Legend* (1935) and Cleveland Amory's staccato treatise which has given the language a label in *The Proper Bostonians* to stand beside John P. Marquand's *The Late George Apley.* As to the last, I suppose that it is in fact the Boston classic of our time.

EARLY AND LATE I suppose that anyone who loves the country must feel first of all that one of the great charms of Boston is that even in her recent decades of incredible expansion she has never lost the look of everlastingness that belongs first of all to the New England town and the village green. I have alluded more than once to the way in which the seasons—all the startling changes locked in the year's portfolio—bring botany and wind and weather right to Brimstone Corner, Harvard Square, and Dorchester Heights. In spite of narrow streets in unexplained directions, a Bostonian never feels shut in, with a barrier of granite on the one hand and the loneliness of anonymity on the other. The tide rolls in and out from the country on three sides, and from the sea on the other, and the people who live here are a part of that tide as they always have been.

I like the way that Boston leans back in brick and stone, in soot and smoke, against her diminutive hills. I like the way the old red face of her peers through the blue mist in the morning and dissolves into the dark when the sun drains out of the solvent streets. I like the slant of light at midday on the thousand chimney pots on the one hand and the smoky granite on the other. I like the white plumes of steam that issue swirling from her downtown rooftops on a cold and blowing November afternoon. I like the gentle look of rain on her doorstep. I like the deep laryngal sound of harbor whistles floating in on the cold damp easterly, just ahead of enveloping fog. I like the fellowship of church bells on a Sunday morning and the spell of chimes and carillon hymn tunes drifting north across the Back Bay. I like the inexhaustible freshness of an October day and the strongest March winds that ever blew in any city east of Michigan. I like the flight of lavender windowpanes on Beacon Street and the two or three on Commonwealth, and the sudden glimpse of red and blue and yellow in the inner gardens back of unexpected windows. I like the occasional iron handrails to help the old and young down slippery streets on slippery days. I like the narrowness of streets that should be wide and the wideness of streets that might be narrow. I like the old-fashioned faded lettering on the signs of shops and warehouses leaning toward the harbor. I like the market set off by Greek New England buildings, and general bedlam, and open carts so openly arrived at, blending the exuberance of fish with the indestructibility of garlic. I like the many steps on Blackstone Street leading down by night to lighted markets underground. I like the multiple names of stores that read like the multiple names of distinguished law firms. I like the patchwork of old and honest iron and brass surviving into a century that greatly prefers a chromium front and the cheap streamline of fake obsidian. I like the austere slant of the Common and the contrasting level radiance of the Public Garden.

I like the placid, imperceptible flow of the Charles—a river that somehow failed to hurry throughout history, and winds to sea from under a bridge named for Longfellow and between the Cambridge and Boston symbols for a bottle of ink and a bottle of iodine. I like, as any youngster likes, the roar of trains into and out of the heart of the city, and the gleaming arteries that stretch away to Ottawa, Seattle, New Orleans, and Mexico. I like to rediscover how endlessly fascinating it is to forsake the noise and trammel and fever of Boston for the lure of the deep canyons of New York and return home pleasantly shocked by the drowsiness of a wise old village I had forgotten.

I like the people of Boston for what they were and are and will be. One has to admire the conservative quality in them that keeps old institutions alive, good reputations intact, and a strong idealism predominant. With the growth of the nation certain things have passed out of New England, industrial and otherwise, for you can't shift the center of gravity anywhere without displacement of its influence. Yet Boston has never lost her universal supremacy for being independent in character, original in enterprise, unwilling to follow whenever she is reasonably equipped to lead. If she has surrendered any of her intellectual heritage, she is still too occupied in serving the humanities and human beings to pause for an audit.

Industrious in industry, she has never stopped

growing. Her natural conservatism has at no time interfered with generous instruction in civil liberties. She has been a proving ground for everything from invention and discovery to politics and law. She is modest. She has more visitors annually than can very well be handled, but let no one tell you that she lacks in hospitality. She has been sketched by thousands but has never had time to sit for a full-length portrait. She is an old city constantly renewed, but she has never had her face lifted. She is part of America, and the spiritual home of many who have never seen her rising from the sea, but whose fathers saw her as theirs before them.